Form, Structure and Mechanis

FORM, STRUCTURE AND MECHANISM

MICHAEL FRENCH

Professor of Engineering Design
Lancaster University

© Michael French 1992

All rights reserved. No reproduction, copy or transmission of this publication may be made without written permission.

No paragraph of this publication may be reproduced, copied or transmitted save with written permission or in accordance with the provisions of the Copyright, Designs and Patents Act 1988, or under the terms of any licence permitting limited copying issued by the Copyright Licensing Agency, 90 Tottenham Court Road, London W1P 9HE.

Any person who does any unauthorised act in relation to this publication may be liable to criminal prosecution and civil claims for damages.

First published 1992 by
MACMILLAN EDUCATION LTD
Houndmills, Basingstoke, Hampshire RG21 2XS
and London
Companies and representatives
throughout the world

ISBN 0-333-51885-6 hardcover
ISBN 0-333-51886-1 paperback

A catalogue record for this book is available from the British Library.

Printed in Hong Kong

Contents

Introduction: the Use of this Book   ix

Acknowledgements   xi

1 Preliminaries   **1**
    1.1   The scope of this book   1
    1.2   The anatomy of design   1
    1.3   The conceptual stage   4
    1.4   General aids to design   4
    1.5   Combinative methods: tables of options   11
    1.6   Repêchage and reviews   13
    1.7   Recurrent problems in design   14
    1.8   Combination and separation of functions   18
    1.9   Summary   22

2 Structures   **23**
    2.1   Introduction   23
    2.2   Structural elements: rods, ties and struts   25
    2.3   Beams   26
    2.4   Plates   31
    2.5   Torsion   31
    2.6   Thin-walled pressure vessels   33
    2.7   Contact stresses   34
    2.8   An example: bathroom scales   38
    2.9   Bathroom scales: refinements and details   41
    2.10 Springs   44
    2.11 Structures, springs, energy and stiffness: pertinacity   48

3 Abutments and Joints — 52
- 3.1 Introduction — 52
- 3.2 The form of abutments — 52
- 3.3 Screwed fastenings — 56
- 3.4 Joints — 60
- 3.5 Stator blade fixing — 62
- 3.6 Joint efficiency — 64
- 3.7 Offset bolted joints: lugs — 68
- 3.8 General principles applied to joint design — 71
- 3.9 Register in joints: intersection problems — 73
- 3.10 Alternating loads on bolted joints — 75
- 3.11 Sealing — 77

4 Freedom and Constraints: Bearings — 79
- 4.1 Degrees of freedom — 79
- 4.2 Shafts and bearings — 81
- 4.3 The principle of least constraint: kinematic design — 83
- 4.4 Epicyclic gears — 84
- 4.5 Bearings — 86
- 4.6 Hydrostatic bearings — 87
- 4.7 Hydrodynamic bearings: squeeze action — 91
- 4.8 Hydrodynamic bearings: wedge action — 94
- 4.9 Rolling-element bearings — 96

5 Various Principles — 100
- 5.1 Insight and abstraction — 100
- 5.2 Biasing — 101
- 5.3 Force paths — 102
- 5.4 A cylinder head joint — 104
- 5.5 Bevel gear mounting — 107
- 5.6 Gear pump — 107
- 5.7 Nesting and stacking — 110
- 5.8 Guiding principles for choosing nesting orders — 112
- 5.9 Other aspects of nesting — 114
- 5.10 Summary of guiding principles for nesting orders — 118
- 5.11 Flexural elements — 118
- 5.12 Easements — 121

6 Materials and Manufacturing Methods — 123
- 6.1 Introduction — 123
- 6.2 Materials, manufacture and design philosophies — 125
- 6.3 The effect of density — 128
- 6.4 Choice of material — 129
- 6.5 Figures of merit — 131

6.6	The relation between form and manufacturing method	137
6.7	New materials and processes	141
6.8	Aids to material selection	142

7 Pneumatic Quarter-turn Actuators — 144
- 7.1 Introduction — 144
- 7.2 Fundamental considerations — 145
- 7.3 Aspect ratio — 148
- 7.4 Table of options — 149
- 7.5 Studying a combination — 152
- 7.6 Two good designs — 156
- 7.7 Summary — 159

8 Epicyclic Gears — 160
- 8.1 The function of the planet carrier — 160
- 8.2 Increasing the stiffness by asymmetry — 163
- 8.3 Local form design — 164
- 8.4 Joint location and design — 167
- 8.5 'Alexandrian' solutions — 167
- 8.6 The choice of embodiment — 169
- 8.7 Another solution — 171
- 8.8 Other aspects of epicyclic gearing — 173

9 Hydraulic Pumps — 175
- 9.1 The swash-plate pump — 175
- 9.2 Valve plate design — 176
- 9.3 Eliminating the valve plate — 179
- 9.4 Virtues and limitations of the swash-plate pump — 179
- 9.5 The bent-axis pump — 180
- 9.6 An elegant bent-axis pump — 182
- 9.7 Combining the virtues of swash-plate and bent-axis — 185

10 Miscellaneous Examples — 191
- 10.1 Connecting rods — 191
- 10.2 A suitcase handle and a suspension arm — 195
- 10.3 A window stay — 198
- 10.4 The *Challenger* disaster — 203

11 The Principles of Design — 208
- 11.1 An emerging discipline — 208
- 11.2 Least constraint and kinematic design — 209
- 11.3 Separation and combination of functions — 210
- 11.4 Design mating surfaces or abutments — 211
- 11.5 Clarity of function — 211

11.6	Short direct force-paths	213
11.7	Matching and disposition	214
11.8	Nesting order and related principles	216
11.9	Avoiding arbitrary decisions: combining good features	217

Bibliography 220

Index 222

Introduction: the Use of this Book

The chief aim of this book is to help students of mechanical engineering to learn to design and to understand the deep and subtle relations between design and engineering science. It is also an original book, in which there are developed ideas which are not to be found in such a complete form elsewhere, and on that account it will be of interest both to experienced engineers and to those with other interests in design.

Experienced mechanical engineering designers build up characteristic ways of thinking and stocks of ideas, and this book aims to help the student to lay the foundations for such expertise early on and in a superior way. The scope is limited to form, structure and mechanism, because it is these areas which are crucial and in which design thinking is best seen. It is not intended to replace books on machine component design, such as those by Shigley and Juvinall, and it is expected that the reader has or will study such a book.

The student should simply read this book, checking the mathematics where appropriate and noting the engineering science given in condensed form where it is required. The reasoning should be followed with special care, because it embodies the *design thinking* that it is intended to teach.

Valuable ideas spring from insight and a firm grasp of the essence of the problem, so much of the book is concerned to show how to develop these prerequisites in key areas such as structures. This understanding is then applied, in the later chapters, to a number of products. The final chapter summarises and further illustrates the ideas which have been presented in the rest of the book.

It is desirable, but not essential, that the reader should understand simple statics (forces, moments and equilibrium), engineers' bending theory and instantaneous centres. Sixth form calculus is also necessary to develop a few of the results. Some of the ideas depend on an ability to visualise relationships in three dimensions, and here a rough cardboard model may sometimes help.

Experienced and able designers will recognise many of the ideas in this book, and perhaps recall the time it took to acquire them. Such readers, I hope, will find here, together with things they already know, others which will be new and interesting, and sometimes of direct applicability in their work.

I have chosen examples which are typical of a wide range of problems, which are simple, and where the background is not too complicated to explain in a few paragraphs. I have probably overlooked some excellent examples which might have been used, and I should always be grateful to any reader who reminded me of one, or, even better, introduced me to one that I had never met. A beautiful concept in design is like a brilliant development in a string quartet, a source of delight which never fails.

Acknowledgements

My thanks are due to those persons and organisations who have kindly given me permission to use material, as follows:

Professor M.F. Ashby	Figure 6.12
Worcester Controls Ltd	Figures 7.9 and 7.10
Volvo Hydraulics Corporation	Figures 9.6 to 9.8
Mr R.C. Clerk	Figures 9.12 and 9.13
Wärtsilä Diesel	Figure 10.5
Jaguar Cars Ltd	Figure 11.4
Forac Ltd	Cover illustration

I am grateful also to BMW for their kind assistance and their correction of Section 1.7, and to M.B. Widden for his help in preparing Figure 2.9.

I am deeply indebted to the very large number of persons who have helped shape my understanding of design over the years. Rather than attempt to list them, an exercise which would be tedious and bound to result in oversights I should regret, I will just put on record my gratitude to all those who have helped in my education and continue to do so. Just one name I will mention, that of Alex Moulton, who, in the course of many years and largely through a voluminous correspondence, has opened my eyes to a multitude of interesting things.

Last and most importantly, I thank my wife, Helen, for all her help and patience over many months.

1

Preliminaries

1.1 THE SCOPE OF THIS BOOK

This book deals with the choice and refinement of forms, particularly of structure and mechanism. This crucial activity occurs towards the end of the conceptual stage of design and right through the embodiment stage. It forms a large part of the most characteristic work of the designer, and it can be highly inventive and elegant. It bears strongly on both the performance of the product and its cost of manufacture, and it is also a prime determinant in appearance and in the satisfaction of ownership.

Able and experienced designers are good at such work, but they are few and have taken a long time to acquire their skill. Moreover, even they may overlook opportunities for improvement or find themselves faced with unfamiliar problems, and then they may find this book of value, for it gathers together and puts into order many ideas, aids and principles from the whole field of functional design.

The readers to whom this book is chiefly addressed, however, are the engineering student and the young graduate. It introduces them to some of that distinctive kind of thinking which characterises creative engineering and enlivens the rather dull science, the crowning glory of the subject – design.

1.2 THE ANATOMY OF DESIGN

The stages of the design process are shown in Figure 1.1. Many such diagrams have been produced, and many of them are very elaborate, but a

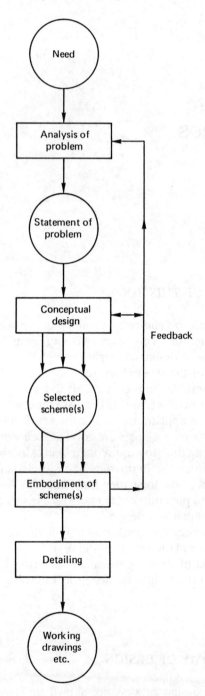

Figure 1.1 Block diagram of design process

simple form is preferable and this particular pattern has been widely followed since the writer adopted it in 1971.

The process begins with a need – either an expressed need, from a client seeking tenders, or a known and conscious need, such as those for cars and cookers, or a need, unperceived as yet by the market, which some entrepreneur seeks to meet. The first step is to analyse the problem, collect relevant data and produce a *statement of the problem*. Such a statement is sometimes called a 'design brief' or a 'specification', and in different cases one or other of these terms may be more appropriate, but 'statement of the problem' is a good generic name. This statement should normally have three parts: the problem itself, the constraints placed on the solution, such as statutory requirements, date of completion, national standards and so on, and the criterion of excellence which is to be used in judging the quality of the design. Frequently the last is simply that it should offer better value for money than some earlier or rival product.

The core of the statement, the problem itself, must list all the functions which are required of the design, together with features which are desirable, rather than essential, ideally with some allowable cost attached to these 'desiderata'. Thus it is a desirable feature in a small lawn-mower that it should have the height of cut adjustable from the handlebars, but whether such a feature is provided in a given model will depend upon how much it adds to the price and informed judgement as to what the market will stand.

Note that the feedbacks shown in Figure 1.1 go back as far as the analysis of the problem. This is because work done in the later stages often causes our view of the problem to change, sometimes radically.

The analysis of the problem requires many specialist inputs. It may, for instance, require that market surveys are made to determine whether there is likely to be a demand for the product or what features it should offer. It may require careful studies of the financial implications of embarking on the new product, and of other business aspects such as the capacity of the production facilities and the relation of the new design to the general product mix, the outlets through which it would be sold, the provision of servicing and spares, the 'launch', and so on. But many of the questions essential to this stage can only be answered properly when we know more about what the designers can achieve, for example, can it really be made for less than £40, will one man be able to operate it, how much extra will it cost to increase the servicing interval to a year, can it be made principally from pressings in-house, or will it have lots of castings we must buy in? Even the analysis of the problem forms part of the iterative procedure which is so characteristic of design, where the chicken-and-egg problem is rife.

However, it is not the purpose of this book to dwell on these peripheral aspects, which have already received plenty of attention elsewhere, and are largely matters of commonsense anyway. The next stage, that of conceptual design, is where the peculiarly designerly aspects begin to emerge, including those with which this book is mostly concerned.

1.3 THE CONCEPTUAL STAGE

The conceptual stage generates broad solutions to the problem in the form of *schemes*, which may be defined as follows.

> "By a *scheme* is meant an outline solution to a design problem, carried to a point where the means of performing each major function has been fixed, as have the spatial and structural relationships of the principal components. A scheme should have been sufficiently worked out in detail for it to be possible to supply approximate costs, weights and overall dimensions, and the feasibility should have been assured as far as circumstances allow. A scheme should be relatively explicit about special features or components but need not go into much detail over established practice."
>
> (M. J. French, *Engineering Design: The Conceptual Stage*, 1971).

To help us in the conceptual stage there are a number of techniques and approaches which will be treated later, and also some very general procedures and attitudes which predispose to success.

1.4 GENERAL AIDS TO DESIGN

(a) Develop insight

Especially with unfamiliar problems, it is important to develop insight into them, and to do so as rapidly as possible. For this purpose, many rough sketches and simple calculations should be made, covering all aspects. Premature detailed studies may prove to have been a complete waste of time when some hitherto unconsidered facet of the problem is studied, and shows that a radical revision of ideas is necessary.

Discussion with others often leads to flashes of insight. It has been suggested that inventive ideas come after a period of 'incubation'. When we return to a problem after a spell working on something else, then very often we see things in a different way, often more clearly. Examples of the development of insight occur in the problems later in this book.

(b) Diversify the approach

When a problem proves difficult, it often helps to try a different line of attack. A good example is provided by the domestic water tap, where it is difficult to ensure that it cuts off completely. The traditional approach is to use a soft washer which is squashed down on to the valve seat: any imperfections in the mating surfaces are taken up by the deformation of the soft material, providing a good seal. However, it is possible to take a different line altogether and make the mating surfaces both very hard and very smooth and flat. This approach works well if the surfaces are hard enough and resistant enough not to be worn by the water, and recently excellent taps have been made on this principle.

(c) Proceed stepwise

This injunction may seem unnecessary, but it is easy to overlook the fact that most ideas do not arrive in their final form and so reject what might be the beginning of a fruitful process. Crossing a mountain stream in the dark one might hunt for a stone just a little way out, and then from there another, and so on. When the daylight comes it is possible to see the whole way across and the difficulty of the night is gone. Faced by a seemingly intractable problem, collect ideas which might prove useful and try to piece them together. If you can, find ways round their limitations. Here is an example of such an approach.

An example of stepwise progress

A student produced a fundamentally unsound idea as a way of centering discs on shafts in high-speed rotors, where centrifugal loads stretch the discs so that their clearance on the shaft increases often by as much as 0.003 times the shaft diameter. If no means were provided of keeping the discs central, no *centering means*, the rotor would go disastrously out of balance.

The problem may be circumvented by large interference fits, making the hole in the disc smaller than the shaft and heating the disc to get it on, or by flank-fitting splines (see Figure 1.2) which fit even when the disc grows relative to the shaft, or by various other means.

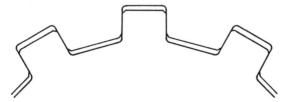

Figure 1.2 Flank fitting splines

The student's idea is shown in Figure 1.3a. His thinking was, that the mercury, in reservoirs in the disc, would be raised to a high presure by the centrifugal field and would press the three pistons hard against the shaft, taking up the clearance produced by the growth of the disc.

There are two things wrong with this idea: the mercury is just flung outwards by the rotation, and so would not press on the pistons; the pistons, even if they were forced inwards, would remove the clearance but would still leave the disc free to move off-centre. Now it might be a reasonable reaction to forget all about an idea so very seriously flawed, but it will be seen how it led to a good solution by a series of steps.

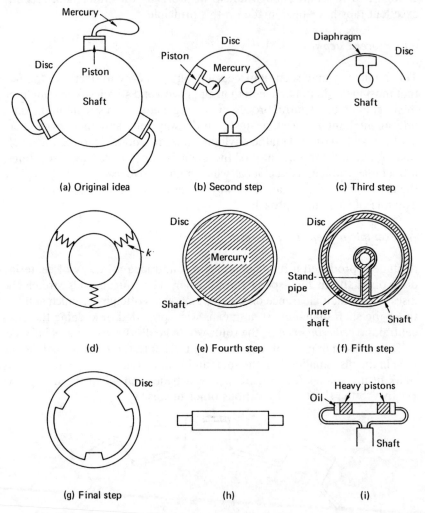

Figure 1.3 Stepwise invention – centering discs on a shaft

The first step is to put the mercury inboard of the pistons, so they would be forced outwards (Figure 1.3b). The next step overcomes the lack of centering action (Figure 1.3c): the 'pistons' are replaced by flexible diaphragms in the shaft surface. It is still *kinematically* possible to move the disc off-centre, but if the diaphragms are stiff it requires a large force to do so. The mercury itself produces a negative stiffness: if a diaphragm moves outwards so does the mercury behind it, which will increase the pressure in it. The effect of the stiff diaphragms is shown in Figure 1.3d, where the stiffness k of a spring is the stiffness of a diaphragm less a small negative stiffness due to the mercury. The overall stiffness of the mounting of the disc on the shaft is $1.5k$.

However, the manufacturing problems alone of Figure 1.3c are enough to make us look further. Concentration on overcoming the formidable difficulties of the diaphragms leads to the elegant simplicity of Figure 1.3e, a scheme which is workable but very heavy, extremely expensive and potentially dangerous. It also means that no use can be made of the bore of the shaft, for example, as a duct or for an inner shaft.

Seizing the essence

It is time to take stock. A good guiding principle in design, as in much else, is to try to seize the essence of the matter, which is here, to cause the shaft to expand under centrifugal effects equally with the disc. In Figure 1.3e, we do not need all that mercury, only the pressure it exerts on the bore of the shaft, and that can be achieved by means of a thin layer of mercury pressurised by a relatively small column of mercury, a stand-pipe in effect, as shown in Figure 1.3f. However, the manufacturing problems are serious and the use of the bore is still lost.

One principle of design is not to make arbitrary decisions, or rather, since it is scarcely possible to start work without making some decisions, to review them at an early stage: even in this form, the principle is difficult to follow. To stretch the shaft *uniformly* as much as the disc, as is done in Figure 1.3f, requires an enormous force. Now the schemes of Figure 1.3e and f conceal an arbitrary decision, which is, to stretch the shaft uniformly. This is a common kind of unperceived arbitrary decision, to assume uniformity or symmetry without even questioning whether it is appropriate or not. In this case, a radially-symmetrical distribution of force is the least suitable for stretching the shaft to maintain the fit of the discs upon it. Forces concentrated at two diametrically-opposite points will stretch the shaft into an elliptical form, producing the same diametral growth as a uniformly distributed force perhaps a hundred times greater. In the uniform case the shaft wall must be stretched, but with local forces it need only be bent.

Now an elliptical deformation would not answer the problem, because it would not prevent the disc moving off-centre along the short axis of the ellipse. The least number of high points that will ensure centering is three, as in Figure 1.3d. The force needed for a given growth is larger, but still very much smaller than for the uniform loading.

A solution

The easiest way to produce this now quite modest force is by means of lumps on the inside of the shaft bore, as in Figure 1.3g. In the case studied, the shaft diameter was about 100 mm and the 'lumps' needed to be each about 10 mm square. Notice that the material of the lumps is not wasted structurally: it is not much use for transmitting torque, but it does add to the second moment of area of section and hence to the bending stiffness of the shaft nearly as much as if it were part of the wall.

Manufacture is difficult if the shaft has to be smaller in section at the ends, as in Figure 1.3h, especially as it is usually important to keep the balance of the shaft good. One solution is to make the shaft in three sections, a long uniform centre section with two ends brazed in in a furnace.

Two steps have been omitted. In Figure 1.3f, it would be possible to use oil pressurised by heavy pistons, as in Figure 1.3i, where everything shown is rotating solidly with the shaft. Notice that two pistons are needed to maintain balance, and also that the system as shown is unstable – if one piston moves out, the force on it will be greater, and the other piston will be pushed in, and so on. Consider what means could be used to remove the instability.

Starting with a totally unworkable idea, a series of more-or-less natural steps led to an elegant and simple solution. The morals are 'even a bad idea may lead to a good design', and 'take small steps, but keep moving until you arrive'.

Before leaving this long example, note that in most cases some feature would be necessary to prevent the discs turning on the shaft, such as a few splines, and the centering is not very stiff, that is, it is rather like that of Figure 1.3d, so that this device might be unsuitable where transverse stiffness is critical.

(d) Increase the level of abstraction

Perhaps the most useful single general aid to the designer is the abstract approach, looking at the problem in a very abstract form, divesting it of circumstantial details which may confine us in mental ruts. The French philosopher Souriau wrote "*Pour inventer, il faut penser à côté*" (to invent, one must think to the side), but this is a misleading image, along with divergent/convergent thinking. In design problems, there is often no set way to diverge from, and when we look back on a brilliant piece of design with hindsight, it seems direct, rather than circuitous. That we cannot see these things is often because we are standing too close, we cannot see the wood for the trees. If we step back, if we take a more abstract view, we can see the wood.

The example of disc centering included a piece of 'stepping back', of increasing abstraction. After the solution of Figure 1.3e, we seized the essence of the matter, the abstract description of what we were trying to do, which was to cause the shaft to expand equally with the disc. We then saw the decision to expand it uniformly was an arbitrary one, and indeed, one that made the task as difficult as possible.

A historic example

Another reason for adopting an abstract view is that many more precedents in design become relevant and may be of help. Figure 1.4a shows an old toy in which a bird is seen in a cage. The bird is painted on a piece of card and the bars of the cage are made by cutting slots in another card placed over the top, the spaces being equal to the bars. The bird is painted on the back card only where it is visible between the bars. Sliding the bars to the side by their own width conceals this bird, and reveals another bird, facing the other way, painted on the strips which were concealed. By moving the bars quickly, the bird appears to jump around.

In the declining years of the reciprocating steam engine early in this century, when it was being displaced by the steam turbine and the internal combustion engine, great ingenuity was deployed in attempts to extend its competitive life. The most successful line was the uniflow engine, in which

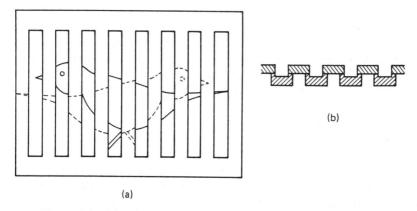

Figure 1.4 (a) Principle of bird-in-cage toy. (b) Section of valve

the piston came nearly to the top of the long cylinder, so that the steam could be expanded to a low pressure in a single cylinder, making for a simple but moderately efficient machine. However, it was crucial to admit the steam during the very short time in which the piston was practically stationary at the end of the in-stroke, and this in turn required an inlet valve of large area opening and closing very fast. One design resembled the bird-in-a-cage toy, with the bars constituting the moving part of the valve and the left-facing bird the valve opening. The right-facing bird was another set of bars, in effect, so that the valve was open when the left-facing bird was visible and shut when the right-facing one was seen. Because the movement was small it could be made very quickly.

Because of the need to seal the valve, the bars had to be slightly wider than the slots and surrounded by a sealing edge, as in the section shown in Figure 1.4b.

It is not suggested that the design of the valve was prompted by the toy, but that both illustrate the same principle, that a large area can be opened or closed rapidly by shaping it as a long narrow slit, which in both these examples is divided up into short lengths arranged in a rectangle.

Valves in internal combustion engines

For an example of the more general or abstract form of the principle, modern car engines often have two inlet valves of the usual poppet type instead of one. Even a single poppet valve has an opening in the form of a slit, whose length is the circumference of the head and whose width is the lift. The 'aspect ratio' of the slip, its length divided by its width, can be effectively doubled by using two valves instead of one. This is a great help in obtaining sufficient opening in a high-speed engine; it improves the 'breathing'.

It is also worth duplicating the exhaust valves, although the advantage is less, and some engines have as many as five valves per cylinder. Interestingly enough, multiple inlet and exhaust valves were used many years ago in big marine diesel engines running at very low speed, but this time because it was difficult to make, operate and maintain large valves. In those days the engines were very complex four-strokes with extra valves for running in reverse and for starting, and with double-acting pistons. Now they are relatively very simple single-acting two-strokes, but of very refined design, turbosupercharged and with higher efficiencies.

Another design principle emerges on considering 'breathing' in engines very abstractly. It is the intermittent nature of the motion, not the speed of motion, that limits the power output of a reciprocating engine. If we want small engines of very high output, then we must choose steady motion, as in the turbine. In the breathing of animals, the same principle appears to some extent. We mere mammals have a simple 'tidal flow' system of breathing, but birds, which need more power per unit weight in order to fly, have a tidal flow with a superimposed steady flow, a surprisingly complex system. The nautilus, a mollusc having some similarity to the squids but with a large shell, also has a 'steady flow' breathing system, and this enables it to survive in water containing very little oxygen.

Many attempts have been made to free the internal combustion engine of the limitation of reciprocating masses, by designs which involve only steady rotational motion. The most successful has been the Wankel engine, but even this has limitations which severely restrict the fields in which it can compete with the conventional piston and cylinder arrangement. Nevertheless, the Wankel has demonstrated the advantages conferred by steady rotational motion and the challenge to the designer is still there.

Besides these four general aids to design – developing insight, diversifying the approach, proceeding stepwise and increasing the level of abstraction – it would be possible to list more, but the overlaps would become greater and the justification less. In addition to general aids, there are design principles (we have seen one, and more will emerge) and systematic methods.

1.5 COMBINATIVE METHODS: TABLES OF OPTIONS

Various methods have been proposed for designing in a systematic fashion based on some combinative table. Some such methods are called morphological, following their originator Zwicky, but for the most part modern versions are not really morphological at all. A general and descriptive term is 'combinative', because the central feature is a table of options from which various combinations are selected.

The usual procedure is to decide all the functions which must be performed in the thing designed, and to list them down the left-hand side of the table we are constructing. Opposite each function we set down all the means by which it might be done. For example, in a tin-opener we might have 'cut tin', 'guide cutter', 'drive cutter' and 'provide power', four major functions. We might cut the tin by a blade or by a sharp disc. The guiding might be by hand and eye, or by the rim of the tin, or by a radius arm. We might drive the cutter directly by hand, or use toothed wheels climbing along the rim, or we might use a lever pivoting in the centre of the end of the tin. There are really only two practical power sources, the human hand or the electric motor. These possibilities have been set out in Figure 1.5a. Any combination of one means from each line constitutes an embryo design, for example, the combination 'knife, rim, wheels, hand', indicated by a broken line in the figure, is a common type shown in Figure 1.5b. Figure 1.5c shows a type which is 'knife, radius arm, lever, hand'.

Function	Means		
Cut	Knife	Wheel	
Guide	Hand	Rim	Radius arm
Drive	Hand	Toothed wheel(s)	Lever
Power	Hand	Motor	

(a)

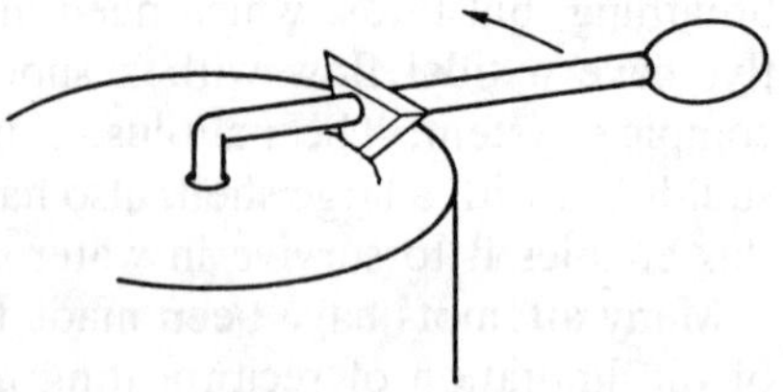

(c)

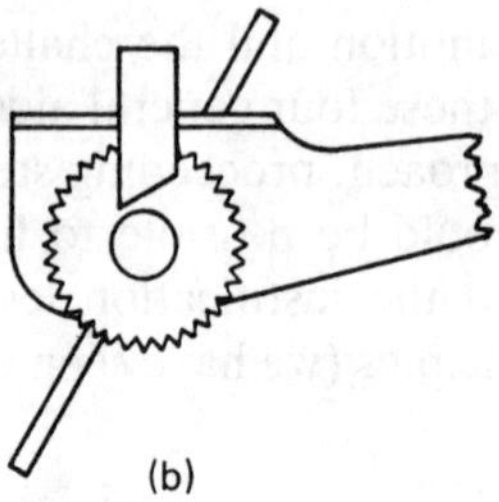

(b)

Figure 1.5   Combinative study of a tin-opener

Some combinations may not make sense: for example, guide by radius arm and drive by wheels do not seem to go together.

This is a modest little table of options for a small design problem. It offers $2 \times 3 \times 3 \times 2 = 36$ combinations, rather less if we eliminate some which appear absurd. We could add other functions: the most important, perhaps, are 'provide reaction', 'start cut' and 'hold cutter into tin', and if we were to add these, difficulties in definition would appear. It is not wise to expect too much of a table of options. Some choices of means add further functions, and some choices automatically supply more than one means. For example, if we choose wheels for the drive we virtually have our guiding means as well. These rather academic defects are no obstacle to the use of the method, unless perhaps if we want to build it into a computer aid.

It is also difficult to limit the lines to functions. For example, the 'wheel' drive means can have the axes of the wheels horizontal or vertical, and we can have one or two wheels. These are differences of *configuration*, and it is convenient to treat these as 'lines', just as if they were functions.

The combinatorial explosion

Larger problems produce more lines, and often, more means. The number of combinations explodes into thousands or millions. The Germans draw tables out on large sheets, with each means illustrated by a tiny drawing, so that the eye can rove over them visualising different combinations.

A naive approach to a table of options would be to choose the best means in each line and combine them. Unfortunately, the *interactions* between functions prevent this system working except in trivial cases. The choice of means in one line affects which choice is best in another line. However, it is clearly impracticable to work out the designs corresponding to every combination and then compare them; there are far too many. More economical approaches must be found.

Condensation and kernel tables

Often there are functions for which one means is clearly the best, or even the only practical one, whatever the rest of the combination. This is not quite the case with the table for the tin-opener (Figure 1.5a), but nevertheless, it is very likely that the guiding means will be the rim. Also, there are often very strong reactions between two lines, so that a decision on means in one will point strongly to a particular choice of means in the other, regardless of the choices in other lines. In such cases it is best to amalgamate or *condense* the two lines into one.

In the tin-opener case, the second and third lines may be condensed in this way, so they form a single line which reads:

> guide and drive: hand and hand, rim and hand, rim and toothed wheel(s), radius arm and lever.

None of the other combinations in the original pair of lines looks practical, so we now have a three-line table with only $2 \times 4 \times 2 = 16$ combinations. Care must be taken, however. Sometimes the best combinations appear unlikely on first consideration, and might easily be thrown out in the course of some over-zealous condensation.

The most effective way of using tables of options may be to select a very few lines and a restricted number of favourable means to construct a manageable *kernel* table. The lines chosen should be those judged to be most critical, and so the outcome will depend on the experience and ability of the designer and his insight into the particular problem. Two simple ways to go wrong are to decide that a function has only one suitable means, and to decide that a function is not of great importance. A great deal of the value of systematic approaches lies in the fact that they ensure that we only make such mistakes deliberately and after due consideration.

1.6 REPÊCHAGE AND REVIEWS

A safeguard against such mistakes is to revise decisions when the work has advanced a little further. To proceed at all it is necessary to make decisions, and to do so before insight into the problem is well developed. The designer should always go back and review such decisions, especially in the early stages. Gifted designers in particular may be carried away with enthusiasm and go into interesting aspects in great depth, finding elegant improvements to one part and developing a strong personal attachment to certain features of their work. Frequent standing back from the task and reflecting on it will help to avoid such effects.

In the author's experience, designers do not generally record their thinking, and this is a mistake, particularly if work has to be dropped and then picked up again later. Every interesting idea, whether it seems good, bad or indifferent, should be recorded and reviewed later.

An example of this kind of practice is repêchage. Suppose a table of options has been reduced to a kernel table and a few combinations from that table have been chosen for study. Often time and resources will not permit all of these few to be taken further, and a single combination is chosen to pursue to the scheme stage (see Section 1.3). We cannot afford to work up all the handful of attractive combinations to this level, but we can return with the greater insight we have developed in elaborating our first choice to see whether perhaps we think another choice might have been better. Such a repêchage is some safeguard against rejecting a superior combination.

1.7 RECURRENT PROBLEMS IN DESIGN

Besides these broad approaches and techniques in design, there are more specialised ones. In particular, there are recurrent situations which come up in different guises in many fields, and similar methods may be adopted every time.

(a) *Disposition*

One of these, called here *disposition*, arises when some commodity, which is usually space but equally well may be strength, temperature difference or bandwidth, is in short supply, and the best possible use must be made of it. It must be shared out between various *functions* to the greatest advantage.

A classic example of a disposition problem arose in small car design. Here the commodity was space, always at a premium in cars and particularly so in small ones. When the engine was disposed in what used to be the traditional way, with the crankshaft running fore and aft, there was a space on either side of it which was little used and of little use (AA, Figure 1.6). By putting the engine transversely, as shown by the broken line, extra space was freed at BB, adjacent to the passenger space, where it

PRELIMINARIES 15

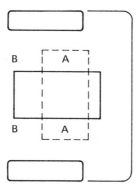

Figure 1.6 Transverse engine

was of great value. Alec Issigonis introduced this arrangement in the Mini, and it was an immediate success: now almost all small cars are designed on these lines.

An interesting example of a disposition problem arises in the arrangement of furniture in small rooms. A small bed-sitting room is typical. The functions are not just to accommodate chest of drawers, bed, table, chairs and wardrobe, but also to provide access to the bed, chest and wardrobe, opening space for the door, and so on. The problem is complicated by radiators and windows, and serious diseconomies arise if care is not taken, such as dead or nearly dead space in corners (incidentally, this is a worse problem in kitchens).

One of the techniques for reducing problems of disposition is to *overlap functions*, so that, for example, the wardrobe doors swing open over space that also gives access to the side of the bed. In Chapter 7, an engineering example is given of a disposition problem which can be eased by overlapping two functions. Another, very simple, example is provided by the design of 35 mm cameras and their cassettes. In these cassettes, the film is wound on a spool and enclosed in a case which has provision each end for sealing out light. The case of the cassette has a slot, also with a light seal, where the film emerges, but this function is not involved in the disposition problem. Figure 1.7a shows a naive design, in which the functions to be disposed along the axis of the spool have been shown simply placed end to end. At the left-hand end the spool engages with the camera case. At the right-hand end of the spool there is in the camera a dog-clutch that can be pushed in and out to engage a mating part on the end of the spool, providing both the engagement with the camera case and the drive means for winding the film back on to the spool when it has all been exposed. Now the overall length L, as shown in Figure 1.7a, dictates to a large extent one dimension of the camera, and generally the designer will be striving to reduce the size as much as possible. Looking at the figure, it is

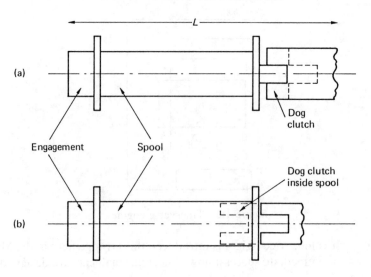

Figure 1.7   Cassette

seen that there is one way of reducing L, which is by overlapping the functions 'spool' and 'dog-clutch', by tucking the dog-clutch into the hollow centre of the spool, thus shortening the stack, as in Figure 1.7b.

(b) Matching

Another common type of problem in design is *matching*. Wind turbines turn at very modest speeds, perhaps 15 rev/min, while electrical generators need to turn at 3000 rev/min, or at least 1500 rev/min. To *match* the wind turbine to the generators needs step-up gearing with a ratio of 1 to 100. A spring needs to be matched to its task, and so on. However, unlike disposition problems, matching problems are not common in the area of design dealt with in this book, although there is an example in Chapter 10.

(c) Intersection

By intersection problems are meant those where two functions demand to occupy the same space, or where the needs of manufacture conflict with those of structural strength. Examples will occur later, but here is an important example, with a very elegant solution due to BMW. Figure 1.8a shows part of the suspension of a front wheel in a car. The wheel is carried on a hub H attached to a wheel carrier supported in bearings at A and B. Steering takes place by rotation of the wheel and carrier about the axis AB. The joints at A and B are ball joints to permit the up and down movement of the suspension as well as steering.

PRELIMINARIES 17

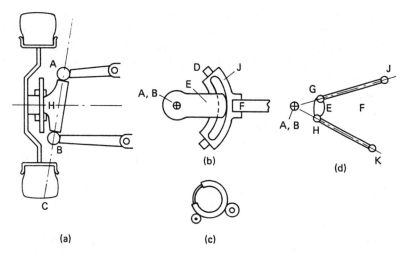

Figure 1.8   Use of remote pivot in steering (BMW)

Now it is desirable that the steering axis AB should intersect the ground near the centre of the contact patch between tyre and road, or if anything, a little outside it in order to improve the control when braking when the friction at the two front wheels is unequal. However, it can be seen that in the figure this intersection point C is considerably inside the centre of the contact patch. It is not possible to correct this by moving both A and B outwards, because other parts prevent this in the case of B. To move A inwards would achieve the required position of C, but only by inclining AB to the vertical by an unacceptable amount. We have an intersection problem because the ideal axis AB passes through other parts which cannot be moved.

The elegant method devised by W. Matschinsky and adopted by BMW to overcome this difficulty was to use a remote pivot, one with no material on or near the axis of rotation. The principle of one such remote pivot is shown in Figure 1.8b, where the view is down the steering axis AB, and D is a bent rod, a piece of a torus, sliding in a toroidal hole in piece J. Both toruses have their centres at B. There is a slot on the upper side of J through which passes an arm E (shown in section) which is rigidly joined to D and forms the bottom end of the wheel carrier. The part J is hinged to the bottom suspension arm (F in Figure 1.8a) to provide the necessary freedom, since the toroidal sliding joint will not do so, although the ball joint at B in Figure 1.8a does.

Now this uncouth solution does overcome the intersection problem. There is a pivot at B, but no material, just an empty space which can accommodate the brake. But the difficulties are considerable. It might be possible to overcome the friction with rollers, overcome the assembly problem by making F a built-up part, provide for effective lubrication and seal the whole thing against grit, water and salt. But a good designer avoids straight sliders if he can, let alone bent ones, which are rare (there is though, the little catch often used to secure necklaces with such a joint, shown in Figure 1.8c).

The good solution adopted by BMW is shown in Figure 1.8d. The bottom suspension arm F is replaced by two links F and F, the axes of which intersect at B but which are jointed to the bottom end E of the wheel-carrier at G and H. The other ends of F and F are jointed to the body at J and K. All these joints, G, H, J and K, are effectively ball joints, like the human shoulder joint. However, for movements in the plane GHJK, this linkage is effectively a four-bar chain, and the instantaneous centre of the 'coupler' E is at B, that is, for small movements E behaves as if it were pivoted at B, which is just what was required. There are rather a lot of joints, four instead of the three for an ordinary suspension arm (like the one discussed in Chapter 10), but they are easy straightforward joints, not like the monstrosity of Figure 1.8b.

1.8 COMBINATION AND SEPARATION OF FUNCTIONS

The combination and separation of functions are two of the most useful resources the designer has. The combination of two functions in one part comes naturally, as it were: the designer is always on the look-out for ways of doing more with less. One very elegant example is provided by the rear suspension on some cars.

Rear suspension

A common form of car rear suspension is based on a trailing arm to support each wheel, as shown diagrammatically in Figure 1.9a. The wheels (not shown) are carried on stub axles at W, W, attached to the L-shaped arms, A, A, which are hinged to the car body along the axis XX. Coil springs S, S between the arms and the body force the wheels downwards. If the suspension were left like this, it would be purely independent, that is, if the car body were inverted and we pushed one wheel down slightly, the wheel on the other side would not move at all.

If we push a car bodily downwards on its suspension, the springs push back upwards with a force proportional to the distance it is pushed down: the suspension has a stiffness k at each wheel, called the *bounce* stiffness (k is generally different for the front wheels and the back wheels, but this will not enter into the argument). If we subject the car to an overturning moment about its longitudinal axis, it will *roll* through an angle θ, and each pair of wheels will resist this roll by a moment $K\theta$, so the pair of wheels

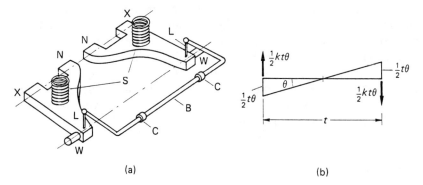

Figure 1.9 Trailing arm suspension with anti-roll bar

have a stiffness in roll of K. If the distance between the wheels, the *track*, is t, then the deflection at the wheels due to the roll θ is $\tfrac{1}{2} t\theta$ at one wheel and $-\tfrac{1}{2} t\theta$ at the other (Figure 1.9b). In a purely independent suspension, the movements of the wheels will produce forces of $\tfrac{1}{2} kt\theta$ and $-\tfrac{1}{2} kt\theta$, constituting a couple of moment $\tfrac{1}{2} kt^2 \theta$ resisting the roll. Hence

$$K = \tfrac{1}{2} kt^2 \tag{1.1}$$

The anti-roll requirement

This shows that if the suspension is purely independent, the roll stiffness is determined by the bounce stiffness and the track. In practice, the designer wants to choose both k and K, and he cannot vary the track to do this. He therefore introduces some dependence, so that if we invert the car body and depress one wheel, the other moves. Normally, if k has been chosen then the value of K given by equation (1.1) is not high enough. To obtain a higher K, dependence must be introduced such that, when one wheel is displaced, the other wheel moves the same way by b times as much, where b is a number less than one. It is left to the reader to show that K is now given by

$$K = \frac{kt^2}{2(1 - b)} \tag{1.2}$$

(*Hint*: apply a force F up on one wheel, a force F down on the other, and note the deflections are $\pm F(1 - b)/k$).

It should be mentioned that these 'inverted car' tests are purely thought experiments. The springs are not linear, so we should also imagine a steady load on the wheels, to bring them into their mean positions, before these 'tests' are applied.

The means adopted to introduce the right amount of dependence is an anti-roll bar, B in Figure 1.9a. This is usually a round rod, secured to the body by two rubber bushes allowing it to rotate, and with two bent-round ends D which are connected by links L to the trailing arms. If one wheel is raised, the link L pushes up the end D, twisting B and applying a force, through the other link L, to the other wheel, tending to raise it also. If the links and the bar B were infinitely stiff, then the two wheels would be forced to move equally, and b in equation (1.2) would be one. By picking a suitable stiffness for B, the designer can choose b.

The combined arms

The combination of functions is done as follows: in Figure 1.9a, the two trailing arms are made into one, by prolonging the parts A and making them into one beam, and the links L and the anti-roll bar are eliminated, to give the arrangement shown in Figure 1.10. If the section A of the combined arm in Figure 1.10 were rigid, the two wheels would be forced to rise and fall as one, and b would be one and K infinite. However, by choosing an open section which has high stiffness in bending but low stiffness in torsion (see Section 2.5), the required value of K can be achieved. The centre section of the combined arms not only carries out the function of the anti-roll bar, it also replaces the inner pair of bearings N, N that the two separate arms required, and what is also valuable, the structure in the body needed to support those bearings. Note that the part A must be strong and stiff in bending, because it will be subject to

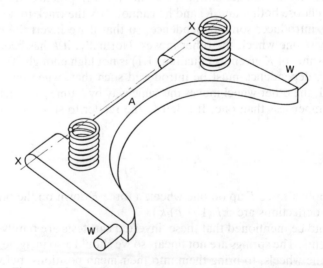

Figure 1.10 Trailing arm suspension with twisting beam

substantial bending moments, and also it must twist considerably without stresses that will lead to fatigue, so that it is not easy to design.

This rear suspension is a very elegant example of combination of functions, and has proved very successful.

Separation of functions

Separation of functions is usually a much less obviously appropriate resource than combination of functions, and yet it is very important. One example might be said to be the division of labour, whose importance was pointed out by the pioneer economist Adam Smith. He drew attention to the extent to which material civilisation had progressed since the tribal period, where each family grew its own food, made its own clothes and built its own dwelling. Much greater productivity is attained where every worker does one job, or one part of one job, like sewing or laying bricks. Separation of functions is just a more general version of the same principle: by providing one part for one function, better performance can be achieved. Occasionally too, more than one part may be provided for one function.

Watt and the condenser

A classical example of separation of functions is provided by Watt's principal improvement to the steam engine. The Newcomen engines that Watt turned his attention to had a cylinder in which steam did work on a piston, and then the steam was condensed by a spray of water so that the piston returned to the bottom of the cylinder again. Thus, in the Newcomen engine, the cylinder functioned both as working cylinder and condenser. Watt found that of the steam supplied to the cylinder in the Newcomen engine, two-thirds was needed just to warm it up again, only one-third doing any work. Watt's great invention was a *separate* condenser which remained always at the lowest temperature that the cylinder had reached before: at the end of the stroke, an exhaust valve opened, admitting the steam in the cylinder to this condenser, into which it rushed. A continuous spray of water in the condenser condensed the steam, exactly as it had been condensed in the Newcomen engine, but the surrounding metal was now always cold, and the cylinder was always hot. No (or strictly, relatively little) steam was wasted in cycling the temperature of the cylinder walls. By separating the functions of working chamber (or cylinder) and condenser, Watt reduced the consumption of coal of the steam engine to about one-third, so vastly increasing the range of applications for which it was economically attractive.

Leaking and bursting

Few examples of the separation of function are as dramatic as Watt's chief improvement to the steam engine. A common pair of functions to separate are those of resisting leaking and bursting. We can carry sand in a fabric sack, because the sack will resist both the leaking and the bursting effects. We cannot carry water in a fabric sack, because it will not resist leaking.

In the advanced gas-cooled nuclear reactor built in Britain (AGRs) a large volume of carbon dioxide at high pressure surrounds the reactor, and is circulated to transfer heat from it to the steam generators or boilers to power the turbines. The reactor, the circulating fans, the steam generators and this large volume of high-pressure gas are all enclosed in a casing, which is prevented from bursting by a multitude of strong steel tendons which form a lattice round the contents, a kind of unwoven basket. The holes in this 'basket' are filled with concrete, and the whole is lined with a thin steel lining. The steel lining prevents leaking: the concrete transmits the pressure on the lining to the steel, which prevents bursting. An alternative would be to combine the prevention of leaking and the prevention of bursting in a thick steel pressure vessel. However, the vessel is so large and the pressure so high that fabricating and welding the great thickness of steel required would be very difficult and expensive, and also the composite construction has great advantages. For instance, it is possible to remove and test the steel tendons a few at a time, to ensure there has been no loss of strength, and the distribution of the load over a large number of tendons means that failure is extremely unlikely.

1.9 SUMMARY

This chapter has analysed the process of design for function and given a brief account of some useful approaches. Various references back to these will occur through the rest of the book, which is devoted to particular aspects of design, those concerned with form, structure and mechanism.

2

Structures

2.1 INTRODUCTION

Most parts of most designs act to some extent as *structure*, they transmit or support loads, and the materials of which they are made and the forms which they are given must be appropriate to this function, as well as meeting many other requirements.

Sometimes the structural function requires only that the part should not break, or should last some minimum time. All that is required is adequate *strength*. In many cases, however, *stiffness* is an important requirement: the dimensions of the part in question must not change under load by more than a small amount.

Suppose we have a casing made of sheet metal 1 mm thick. If we increase the thickness to 2 mm, then both the strength and the stiffness will be increased, but so will the cost and the weight. The designer will therefore generally choose the thickness in an attempt to make the strength and stiffness just adequate. Other things being equal, he will try to use an efficient general form, that is, one which uses the least material to meet the need, and then choose the thickness so that it is just strong enough and stiff enough. In most cases, it will be the strength aspect which is critical.

In practice, the designer will usually decide on the overall form of the parts and then determine the thicknesses to be just sufficient. It is convenient to have some term for all these thicknesses, such as sheet metal gauges, casting wall thicknesses, shaft diameters and so on, and the naval architects and marine engineers use the word 'scantlings'. This word comes from the French *échantillon*, meaning a sample: in the old shipyards they would send to the woodyard a sample of wood of the cross-section wanted to show what was needed. Thus the designer, so far as possible, chooses the scantlings to be just large enough. The basis for doing this is generally to make the calculated stress in the material equal to or less than some chosen value, the *design stress* or *allowable stress*, denoted here by f.

The designer meets the structural needs largely by choosing suitable *forms* and then adjusting the *scantlings* to raise the calculated stresses to f. The ideal is that expressed by Oliver Wendell Holmes in his poem about *The One-Hoss Shay*. This fictional pony chaise was built in such "a logical way" that it lasted exactly one hundred years and then every part failed simultaneously.

For structural purposes, we can regard most forms as built up from a small repertoire of basic structural elements, such as rods, beams and shells. These elements are reviewed briefly in the next few sections: it is assumed that the basic engineering science is known to the reader. A few aspects, particularly contact stresses, are dealt with more fully to bring out points not usually covered in text books but the notes provided are otherwise intended simply as a convenient reminder. Readers who seek fuller information should consult appropriate text books (see Bibliography).

Before proceeding to review the basic structural elements with which the designer works, however, some consideration should be given to the concept of the *design stress* f.

Design stress

The choice of a design stress to which to work is a complex matter, which is dealt with at length in many text books (see Bibliography). It may need to take into account uncertainty in the magnitude of the loads and the properties of the material, combined steady and repeated loads, the effects of inclusions, temperature, the chemical and radiative nature of the environment, and local stress concentrations. The required life, the possibilities of inspection and test during that life, and the consequences of failure need also to be considered. A failure in a car engine is less serious than a failure in an aircraft engine, which is in turn less serious than a total failure in an aircraft wing. A failure in a steel pressure vessel containing a nuclear reactor would be a major disaster, but we could take a few prestressing tendons out of a similar concrete vessel and replace them without serious risk. To cover some of these aspects a larger or smaller factor of safety may be used: alternatively, and more satisfactorily, when the necessary data are available, a statistical approach may be used to ensure, as far as possible, that the risk of failure is acceptably small.

It is frequently useful to reduce a more complex reality to a single equivalent stress. For example, in the case of a material subject to two or three principal stresses, either the maximum shear stress or the shear strain energy criterion may be used (these criteria are named for Tresca and von

STRUCTURES

Mises). Another important case arises when an alternating or intermittent stress is superimposed upon a steady one. In the case of a vehicle suspension arm, for instance, there will be a steady bending moment due to the weight on the wheel, and, superimposed upon that, an alternating moment of variable amplitude caused by bumps in the road.

In such cases, the usual practice is to form a single equivalent steady stress σ_E from the steady stress σ_S and some statistical measure σ_A of the alternating stress, after the fashion

$$\sigma_E = \sigma_S + s\sigma_A < f_S \tag{2.1}$$

where f_S is the steady design stress and s is the ratio of f_S to the alternating design stress. Thus s, which may be about 2.5, is a measure of how much greater the strength is in steady loading than in fatigue, or, if you prefer, how much worse alternating stress is than steady stress. The equivalent steady stress σ_E is just a combination which weights σ_A to reflect this. This equivalent stress embodies the Goodman diagram, in effect: note that in a metal it applies only if σ_S is positive. In the case of the suspension arm, σ_A itself must be a statistical measure because the excitation is irregular in frequency and amplitude.

The determination of f (or f_S) is a long process, and one subject to all kinds of difference of approach. The Bibliography gives references which cover the subject, but for the purposes of this book it is enough to assume this work has been done.

2.2 STRUCTURAL ELEMENTS: RODS, TIES AND STRUTS

The designer has a *repertoire* of means he can use. To sustain loads and maintain parts in the right positions he designs *structures*, which are composed chiefly of simple *structural elements*. The following sections list the chief structural elements in the repertoire and the formulae for stress in them, results which will be mostly familiar but are repeated for a reminder and a convenience.

A *rod* is typically a long, thin element loaded in tension, when it is called a *tie*, or compression, when it is called a *strut* (Figure 2.1). If the cross-sectional area is A and the load in the rod is P, the stress in it is P/A, and the extension is PL/EA.

Ties have the advantage that they are inherently stable, and can be made of a cord, for example. On the other hand, they must have end fixings: struts do not need end fixings, but are inherently unstable. Slender uniform struts with pin-jointed ends buckle elastically under an axial load

$$P = \frac{\pi^2 EI}{L^2} \tag{2.2}$$

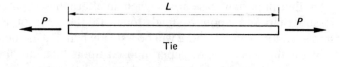

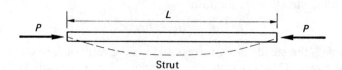

Figure 2.1 Tie and strut

where I is the minimum second moment of area of the cross-section about an axis in its plane and through its centroid (Euler buckling load). A rod which acts sometimes as a tie and sometimes as a strut is inherently unstable and requires end fixings. Some useful additional results on buckling are given at the foot of Figure 2.4.

A design principle which emerges from these considerations is that economical structures tend to have long thin ties (to minimise the cost of end fixings) and short fat struts. Because the loads carried tend to total about the same, this principle leads to many ties and few struts, as exemplified by cable roof structures and tents (Figure 2.2).

Figure 2.2 Simple tent: two struts (poles), many ties (guys)

2.3 BEAMS

A beam is typically a relatively long thin member used in a structure in a way which requires it to resist bending. Figure 2.3 shows some beams

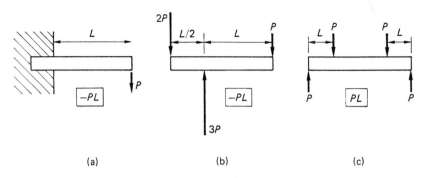

Figure 2.3 Members loaded as beams: extreme value of bending moment shown in rectangles

under load, with the magnitude and location of the maximum bending moment indicated: the negative sign indicates that the moment tends to bend the beam to a shape which is convex upwards (hogging) as against convex downward (sagging). The bending moment M in a beam at a given section can be found by the *method of sections*, that is, by imagining it cut at that section and considering the equilibrium of either of the two resulting parts, having first determined all the external loads and reactions: the shear force F on the section is found in the same way.

Provided the bending moment lies in a plane of symmetry of the beam, the bending stress σ_B at a point distant y from the principal axis is given by

$$\sigma_B = \frac{My}{I} \qquad (2.3)$$

(the principal axis is the axis through the centroid of the section perpendicular to the plane of bending).

When y is a maximum, the quantity $I/y = Z$ is called the modulus of the section, and

$$\sigma_B = M/Z \qquad (2.4)$$

The sections of uniform beams are commonly of simple geometric shapes, for instance, circular, annular, rectangles, angles, I-shaped, T-shaped, Z-shaped, 'top hats' etc. A useful quantity in design is the radius of gyration k, given by $I = Ak^2$. The ratio k/d, where d is the overall depth of the section, is a useful check since it varies little and can be guessed at for a quick calculation. Figure 2.4 shows some typical sections and their properties.

Section	A	I	Z	k/d	Z_p/Z
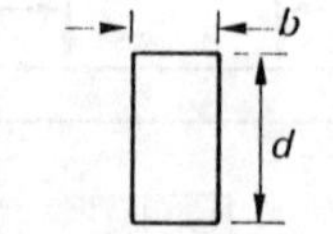 Rectangle ($b \times d$)	bd	$\dfrac{bd^3}{12}$	$\dfrac{bd^2}{6}$	0.29	1.5
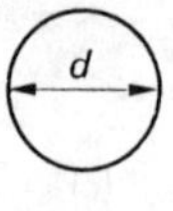 Solid circle (dia. d)	$\dfrac{\pi d^2}{4}$	$\dfrac{\pi d^4}{64}$	$\dfrac{\pi d^3}{32}$	0.25	1.70
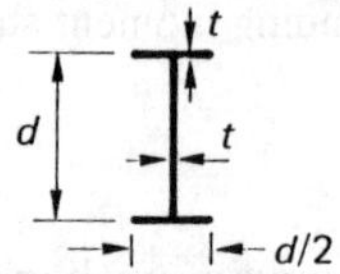 I-section (d, flange $d/2$, thickness t)	$2td$	$\tfrac{1}{3}td^3$	$\tfrac{2}{3}td^2$	0.41	1.13
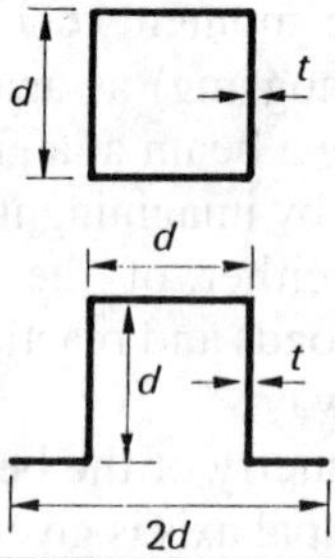 Square hollow ($d \times d$, thickness t)	$4td$	$\tfrac{2}{3}td^3$	$\tfrac{4}{3}td^2$	0.41	1.13
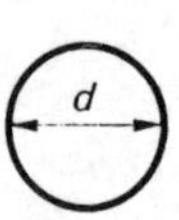 Hollow circle (dia. d, thickness t)	πtd	$\dfrac{\pi}{8}td^3$	$\dfrac{\pi}{4}td^2$	0.35	1.27
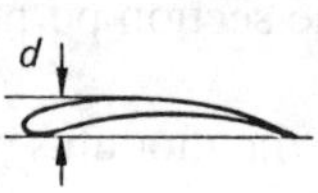 Aerofoil section		$\dfrac{k}{d} \approx 0.23$			

Figure 2.4 Properties of common sections. In hollow sections, $t \ll d$. Z_p = fully plastic moment/yield stress moment

The rationale of the beam: shear: clarity of function

The I-section beam exhibits *clarity of function* very well. The flanges carry bending, with a little assistance from the web, and the web carries shear, with negligible assistance from the flanges. In Figure 2.5, the flanges are of cross-sectional area A_f each and the web is of cross-sectional area A_w, so

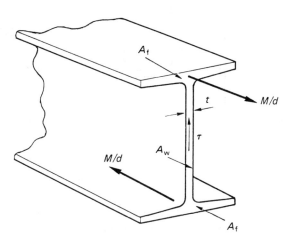

Figure 2.5 Section of beam

the total cross-sectional area is $2A_f + A_w$. The applied bending moment M can be equilibrated by a couple consisting of two opposed forces M/d separated by the depth d between the flanges, so that very roughly the force in each flange is M/d, corresponding to a bending stress σ_B of M/dA_f.

In practice, the material in the web will contribute to resisting the bending moment at about $\frac{1}{3}$ the rate per unit area of cross-section that the flange material does, so a more accurate value is

$$\sigma_B = d(A_f + \tfrac{1}{6} A_w) \qquad (2.5)$$

The web takes the shear force F, and the shear stress is very nearly F/A.

This clarity of function of the I-beam can be achieved in other forms and should be aimed at by the designer whenever relevant and possible. An example is the cylinder head of an internal combustion engine, which to perform its function of loading the gasket round the outside of the cylinder bores to prevent leakage, has to act as a ring beam. This will be discussed in Chapter 5.

It must always be remembered that the shear stresses in the cross-section of the web are balanced by complementary shear stresses along the web (see Figure 2.6a). A useful concept is shear flow q, which is the integral of the shear stress across the thickness of the web or section: it has the dimensions of a stress times a length, that is, a force per unit length. Thus in Figure 2.5, the shear flow q is given by

$$q = \tau t = F/d \qquad (2.6)$$

where τ is the shear stress.

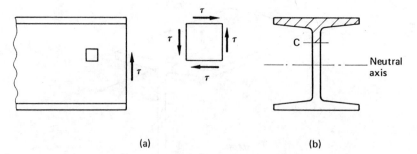

Figure 2.6 Shear stresses in beams

Even in the I-beam, the shear flow q and the shear stress τ vary slightly with depth, according to the familiar formula

$$q = \frac{FA\bar{y}}{I} \tag{2.7}$$

where $A\bar{y}$ is the *first* moment of area about the principal axis of the part of the cross-section above (or below will do equally well, for $A\bar{y}$ is the same) the section C at which q is to be determined (that is, $A\bar{y}$ for the shaded area in Figure 2.6b).

It is important to consider the complementary shear stresses in beams made of anisotropic materials such as wood or plastic laminates, where the shear strength of the material along the grain or the laminae may be much less than transversely.

Deflections of beams

The curvature ρ of a beam is given by the formulae

$$\rho = \frac{M}{EI} \tag{2.8}$$

and

$$\rho = \frac{1}{R} \tag{2.9}$$

where R is the radius of curvature. For straight beams with small deflections, the deflection y of the beam is given by

$$y = \frac{d^2 y}{dx^2} = \frac{M}{EI} \tag{2.10}$$

where x is distance along the beam. By integration and putting in appropriate boundary conditions, many useful results can be found for the uniform beam which has the same flexural stiffness EI throughout its length.

A thin deep beam is very stiff initially in bending, but may buckle and collapse in a way which involves both bending and twisting. The twist produces a bending moment on the beam in its weak plane, which in turn increases the twist, and so on.

2.4 PLATES

Rectangular plates loaded in their plane in one direction only are like wide rods. In compression they are prone to instability and the Euler buckling load formula applies with only the substitution of

$$E' = \frac{E}{1 - \nu^2} \qquad (2.11)$$

for E, where ν is Poisson's ratio, because of the suppression of transverse strain. Stiffeners are often required in thin webs to prevent buckling. Rectangular plates loaded normal to their faces and supported on two edges only are like wide beams, also requiring E' in place of E. Other plates are beyond the scope of this book.

2.5 TORSION

Rods and beams are frequently subject to torsion, and both the stresses and deflections are important to the design. The simplest case is that of a cylindrical rod, of diameter d, subject to a torque Q. The shear stress is proportional to the distance from the centre and so is a maximum at the surface, where

$$\tau = \frac{Qd}{2GK} \qquad (2.12)$$

where G is the shear modulus and

$$K = \frac{\pi d^4}{32} \qquad (2.13)$$

(K is frequently equated with the polar second moment of area of the cross-section, but this is misleading because then the result is true *only* for circular or annular sections: several books contain gross errors from this cause.)

The angular stiffness of a length L of such a rod is

$$\frac{GK}{L} \qquad (2.14)$$

that is, this is the value of Q which would be required to twist it through one radian if it remained elastic. More generally, the twist θ under a general torque Q is given by

$$\theta = \frac{QL}{GK} \qquad (2.15)$$

For a hollow circular cylinder:

$$K = \frac{\pi}{32}(d_1^4 - d_2^4) \qquad (2.16a)$$

where d_1 and d_2 are the outside and inside diameters respectively.

For a rectangular section $b \times d$, when d is very much greater than b:

$$K \simeq \tfrac{1}{3} b^3 d \qquad (2.16b)$$

that is, four times the minimum second moment of area I of the section. When $d = 5b$, this expression overestimates K by 12 per cent, and even when $d = 20b$ the overestimate is still 3 per cent. For an elliptical cross-section, the same approximation to K, that is, 4 times the minimum I, overestimates by only 4 per cent when the major axis d is five times the minor axis b, and only 0.25 per cent when $d = 20b$. For a section like the cambered aerofoil in Figure 2.4, it is necessary to use the minimum I when the camber is removed, that is, the blade section is straightened out.

To estimate the stresses in these flattish rectangles and other forms, the maximum shear stres is given roughly by

$$\tau_{max} = \frac{G\theta b}{2L} \qquad (2.17)$$

This occurs at the surface at the ends of the minor axis.

The resistance to torsion of a flat rectangular section is not increased by bending it into an angle or a channel form, so such forms, like the aerofoil, should be straightened out, as it were, before finding the minimum I.

Open and closed sections

When a form like a channel is twisted, transverse sections do not remain plane but warp out of plane. If the end of a channel is welded to a heavier member all round the periphery, then warping is suppressed and the channel becomes much stiffer and stronger in torsion.

STRUCTURES

If we fold a strip of sheet into a square tube, it is still very weak in torsion as long as the two free edges are not welded together, that is, as long as the tube is not *closed*. Once the two edges are joined, however, warping is suppressed throughout the length and the torsional stiffness and strength increase dramatically, for example, hundreds of times. To convince yourself of this, try twisting one of those long thin square-section cardboard boxes in which kitchen foil comes, first as it is, and secondly with the long opening sealed with tape. In a trial by the writer, the stiffness increased by more than 200 times.

Thin-walled closed sections in torsion are covered by these formulae:

$$q(\text{shear flow}) = \frac{Q}{2S} \quad (2.18)$$

where S is the area enclosed (see Figure 2.7) and

$$\theta = \frac{sQL}{4S^2Gt} \quad (2.19)$$

where t, the wall thickness is uniform (equation (2.18) holds even if t is variable).

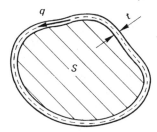

Figure 2.7 Thin-walled tube in torsion. S = hatched area

2.6 THIN-WALLED PRESSURE VESSELS

The hoop stress σ_θ in a thin-walled cylinder of radius r with the internal pressure greater than the pressure outside by p, is given by

$$\sigma_\theta = pr/t \quad (2.20)$$

where t is the wall thickness. The longitudinal stress σ_z is given by

$$\sigma_z = \frac{pr}{2t} \quad (2.21)$$

The stress in the wall of a sphere is

$$\sigma = \frac{pr}{2t} \qquad (2.22)$$

and it is the same in all directions in the surface.

2.7 CONTACT STRESSES

The subjects so far mentioned in this chapter are dealt with adequately in numerous text books. Treatments of contact stresses are generally less satisfactory, or missing altogether, so more space will be given to them and an approach that is particularly appropriate for design will be adopted.

By contact stresses are meant those that occur in a loaded point contact, such as that between a ball and a race in a ball bearing. Only elastic stresses will be considered.

The method adopted will be to consider the contact between an ellipsoid and a plane and then show how to reduce any other contact to that of an equivalent ellipsoid and plane. Finally, an example is worked through.

Contact between an ellipsoid and a plane

Consider an ellipsoid contacting a plane at a point Q. The principal radii of curvature of the ellipsoid Q are R_A and R_B ($R_A > R_B$). When the two are forced together by a load P, the surfaces will deflect until they fit one another over an elliptical contact patch with major and minor axes $2a$ and $2b$ (see Figure 2.8).

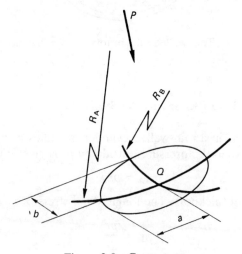

Figure 2.8 Contact stresses

The modulus of elasticity E and the Poisson ratio ν is the same for both components. It can be shown that the relevant modulus in this pattern of stressing is

$$E' = E/(1 - \nu^2) \qquad (2.23)$$

By dimensional analysis we can write

$$a = \alpha \frac{\sigma_{max}}{E'} R_A, \quad b = \beta \frac{\sigma_{max}}{E'} R_B \qquad (2.24)$$

where σ_{max} is the compressive stress at Q. For convenience, in this instance σ_{max} will be regarded as positive, contrary to the usual convention.

In Equations (2.24), the numbers α, β are functions of R_A/R_B only, and they are plotted in Figure 2.9, which was kindly prepared by my colleague M. B. Widden.

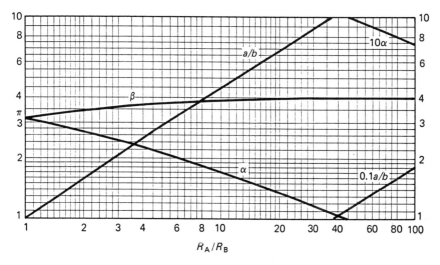

Figure 2.9 Contact stresses: values of α, β (see equations (2.24)). The plot of a/b is often useful in design

It remains to relate σ_{max}, a and b to the load P, for which we require the ratio of the mean compressive stress over the contact area, σ_{mean}, to σ_{max}. The distribution of normal stress over the patch is ellipsoidal, so that

$$\sigma_{mean} = \frac{2}{3} \sigma_{max} \qquad (2.25)$$

Then
$$P = \text{area} \times \sigma_{mean}$$
$$= \pi a b \times \frac{2}{3} \sigma_{max} \qquad (2.26)$$

Relative radii of curvature

In a general contact at a point Q, both surfaces will be curved. However, as far as the stressing is concerned only the *relative* radii of curvature matter. In almost all practical cases, the principal planes of curvature for the two bodies will coincide by reason of symmetries. In this case, if R_{A1}, R_{A2} are the radii of curvature of the bodies 1 and 2 in principal plane A, and R_{B1}, R_{B2} are the corresponding radii of curvature in principal plane B, then the required *relative* radii of curvature R_A, R_B are given by:

$$\frac{1}{R_A} = \frac{1}{R_{A1}} + \frac{1}{R_{A2}}, \quad \frac{1}{R_B} = \frac{1}{R_{B1}} + \frac{1}{R_{B2}} \qquad (2.27)$$

Note that where the intersection of a plane with a surface is concave, the corresponding radius of curvature is negative.

Example of contact stresses

A deep-groove ball bearing has ten balls 8 mm in diameter running on a pitch circle radius of 20 mm. Find the axial load which will give a maximum compressive stress of 1540 N/mm² at the contact with the outer race, given that the dimensioning is such that the common normal at the points of contact intersects the axis of the bearing at 45° (see Figure 2.10). The radius of the cross-section of the outer track is 4.1 mm, and E' is 220 kN/mm².

The first step is to find R_A and R_B. The principal planes are the plane of the figure, which has the biggest relative radius of curvature and so will be called plane A, and the plane perpendicular to it through the common normal.

If we denote the ball by 1, $R_{A1} = R_{B1} = 4$ mm, and $R_{A2} = -4.1$ mm; these values are given. To find R_{B2}, consider the cone generated by rotating OQ about the axis of the bearing. The rim of this cone is tangential to plane B and lies in the surface of the race, and all the points on it are distant QO from the point O. Thus O is the required centre of curvature and

$$OQ = 20 \text{ cosec } 45° + 4 = 32.3 \text{ mm}$$

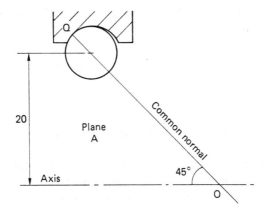

Figure 2.10 Contact stress example

The required radius of curvature R_{B2} is OQ, except that since the surface is concave:

$$R_{B2} = -32.3 \text{ mm}$$

Then

$$\frac{1}{R_A} = \frac{1}{4} - \frac{1}{4.1}, \quad R_A = 164 \text{ mm}$$

$$\frac{1}{R_B} = \frac{1}{4} - \frac{1}{32.3}, \quad R_B = 4.57 \text{ mm}$$

$$R_A/R_B = 35.9$$

From the graph in Figure 2.9:

$$\alpha = 1.1, \quad \beta = 3.95$$

Evaluating a and b from equations (2.24) we find:

$$a = \frac{1.1 \times 1540 \times 164}{220{,}000} = 1.26 \text{ mm}$$

$$b = \frac{3.95 \times 1540 \times 4.57}{220{,}000} = 0.126 \text{ mm}$$

and from equation (2.26):

$$P = \sigma_{\text{mean}} \pi ab$$

$$= \frac{2}{3} \times 1540 \times \pi \times 1.26 \times 0.126 = 512 \text{ N}$$

As there are ten balls and the contact loads act at 45° to the bearing axis, the corresponding axial load is

$$10 \times 0.707 \times 512 = 3620 \text{ N}$$

2.8 AN EXAMPLE: BATHROOM SCALES

So far this chapter has been almost devoid of design: to redress the balance, here is a simple development of some of these ideas in connection with a bathroom weighing machine. The conceptual stage has been completed, and the essential component is to be a spring upon which the user stands and which deflects an amount proportional to his or her weight. The form of spring envisaged is that shown in Figure 2.11a, consisting of two parallel leaf springs rigidly attached at their ends to two blocks B1 and B2.

Why is this a very suitable form of spring?

Firstly, the deflection is nearly independent of where the user stands on the platform. This we shall look at again later.

Secondly, there is no friction, backlash or lost motion to cause error.

Thirdly, it fits well into the available space.

Fourthly, it is cheap.

What are the desired properties of the spring which affect its form?

(1) It must not be overstressed under the maximum load P, which will be taken as 1.25 kN.
(2) It must be insensitive enough to the position of P on the platform: any change in deflection is not to be more than 1 per cent.
(3) The deflection must be large enough to be indicated to an accuracy of, say, $\frac{1}{2}$ per cent of the full scale value using some simple, inexpensive mechanism.

Let us start by taking provisional values for L and b, the half-length and breadth respectively of the leaf springs, as large as can conveniently be accommodated, and calculating a suitable thickness d.

Figure 2.11b shows a section of the scales with the load P over the centre of the springs. By sectioning on LMNO and considering the equilibrium of the upper part (Figure 2.11c) it is seen that there is a shear force $P/2$ and zero bending moment at the centre of each leaf spring. Each half spring can thus be taken as a cantilever of length L with a load of $P/2$ at the tip.

Take $L = 80$ mm, $b = 120$ mm, which seem good first guesses, and set the design stress for these steel springs as 500 N/mm^2: then the bending moment at the fixing G is

$$0.5P \times L = 0.5 \times 1.25 \text{ kN} \times 80 \text{ mm} = 50 \text{ Nm}$$

STRUCTURES

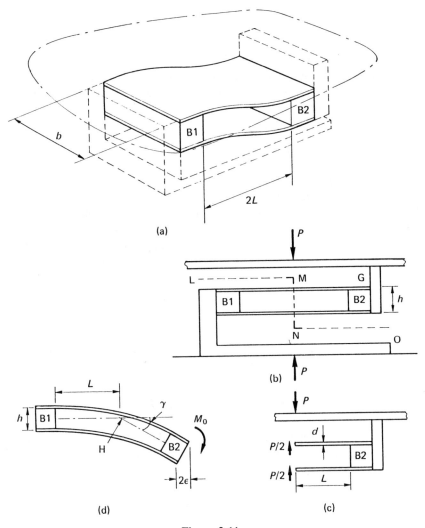

Figure 2.11

To keep the stress to 500 N/mm² the section modulus of the spring must be

$$Z = \frac{M}{f} = \frac{50 \times 10^3}{500} = 100 \text{ mm}^3$$

The section modulus Z is $bd^2/6$ (Figure 2.4) and putting $b = 120$ mm gives $d = 2.24$ mm.

Now calculate the deflection δ, using the expression for a uniform cantilever of length L with a load P at its tip:

$$\delta = \frac{PL^3}{3EI}$$

taking $E = 200$ kN/mm². This gives $\delta = 9.5$ mm, or since there is an equal contribution from the other half of the springs, a total travel of 19 mm. To measure this to 1 part in 200 requires an indicator which is sensitive and repeatable to 0.10 mm, which is a reasonable target.

Effects of a non-central load

Let us check the requirement that the effect of not standing centrally on the platform should not exceed 1 per cent. The limit of standing off-centre is set by tipping up of the scales, and might be put at 130 mm, depending on the overall size of the machine. Suppose the user stands 130 mm to the rear, that is, to the right of centre in Figure 2.11: the effect of moving P to the right is simply to apply to the entire spring an additional clockwise moment

$$M_0 = 130 \text{ mm} \times P$$

The effect of this moment is to bend the spring by stretching the top 'flange' and compressing the bottom 'flange' with forces each of M/h, where h is the distance between the two parallel leaves. Each leaf will have a stress in it of $\pm M_0/hbd$, and hence will stretch or contract by

$$\epsilon = (M_0/hbd) \times 2L/E$$

so the top of B2 will move to the right by ϵ and the bottom to the left by ϵ (see Figure 2.11d), rotating B2 through an angle γ, where

$$\gamma = \frac{2\epsilon}{h} = \frac{4M_0 L}{Ebdh^2}$$

For our design, taking $P = 1.25$ kN and setting h at 30 mm, the force in each leaf, M_0/h, works out at 5.4 kN, the magnitude of the stress in each leaf at 20 N/mm², ϵ at 0.016 mm, and γ at 1×10^{-3} radians. Looking at Figure 2.11d, from the geometry of the centre lines of the fixed block B1 and the moving block B2, which intersect at H, midway along the springs, there is an additional deflection at the centre of B2 of rather more than

$$\gamma L = 1 \times 10^{-3} \times 80 \text{ mm} \quad \text{or} \quad 0.08 \text{ mm}$$

This is in a total travel of 19 mm, and so is acceptably small (in fact, with the mechanism for indicating deflection at H, which will be shown is the best place, the effect is reduced to one half and changed in sign).

STRUCTURES

The virtues of rough calculations

It is worthwhile working out all these intermediate results on the first run through as has been done here, for it gives us some feel for the values and enables us to exercise judgement. It may also remind us that we need to check for buckling in the leaf which is in compression. For a strut with both ends encastré, the value of the constant in equation (2.2) becomes $4\pi^2$, but we must remember the length is $2L$, not L, in the present notation: the buckling load works out at 34 kN, against the 5.1 kN we are applying, which is comfortable, but not very comfortable. Notice that because we worked out the stress at 20 N/mm^2, we know we are only concerned with elastic buckling.

We need also to look at the effect of standing to one side, which tends to twist the spring. The calculations are similar but rather more difficult, but notice that the effect of a twist on the reading will be very small if we measure the deflection on the centreline of the spring unit, which is what we would tend to do anyway. This reminds us to look generally for the ideal point to which to connect the indicating system, which we shall do in the next section.

Note in passing that the leaf springs have a very low aspect ratio, $2L/b = 1.29$, and so the stiffness in torsion will be high because of the suppression of warping by the clamped ends. Twisting is resisted almost entirely by the leaves acting as beams in their own plane, in which mode they are very stiff.

2.9 BATHROOM SCALES: REFINEMENTS AND DETAILS

We started this example by taking some guesses at L and b which could easily be accommodated and a design stress of 500 N/mm^2, and went on to calculate the thickness of spring needed to keep within the design stress. The full scale deflection of 19 mm seemed suitable, although this would depend somewhat on the indicating system, a decision which belongs to the conceptual stage. Let us suppose this is to be a mechanical system: we look at existing mechanical systems for indicating a small movement (the repertoire) and the rack and pinion system used in dial test indicators comes to mind. This is well able to show a 0.01 mm movement, so that a cruder version should be able to show 0.1 mm.

One refinement is prompted by regarding the spring unit as an energy storage means. It is inefficient because the 'cantilevers' in Figure 2.11c are only fully stressed at the root where the bending moment is a maximum. If the cantilever were tapered to its tip, as in Figure 2.12a, the bending stresses would be the same everywhere, and the curvature would be the same everywhere. The deflection for the same root section would be

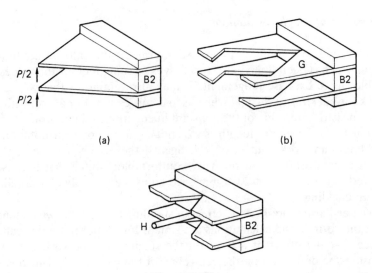

Figure 2.12

increased by 1.5 times. Now the tapered form is clearly impractical, but that shown in Figure 2.12b is workable and will show an increase of deflection of about 1.4 times. The diamond shaped space formed is also ideal for locating the indicating mechanism.

This improvement can be seen as an example of clarity of function. The spring material should all be working fully (as fully as possible in bending, that is, where the efficiency of use is only one-third) and by arranging to approach this ideal more closely we have the bonus factor 1.4 and the useful extra space. What can we do with the 1.4 factor?

It is a general principle of design that in such a case we should study all the uses that a bonus can be put to. A good example is load-alleviation in aircraft wings, which is a system which makes small movements of the control surfaces, independently of the pilot, which reduce the peak structural loading in gusts and manoeuvres. This bonus could be used to reduce the strength of the wing, so saving weight and increasing payload. It could be used to make the wing longer and thinner at the same weight, and so reduce the induced drag, saving fuel and thereby also increasing payload. It could be used partly in each of these ways.

In the case of the scales, we could keep the same maximum deflection and reduce L and b, thus saving material and other costs and making the machine more compact and lighter. We could just make the springs thicker by about 12 per cent, so reducing the stresses by about 20 per cent, and enabling us to use cheaper steel, or use the same steel and increase the load capacity.

There are two more points before we leave this example. The best place to locate the input to the indicator, for example, the rack of a rack-and-pinion system, is at H in Figure 2.11d. This reduces errors due to standing off-centre (not that this matters much, as our calculations have shown), as H is near the centre of the machine and there is ample space round it. Figure 2.13 shows a suitable indicator mechanism, with bearings, the scale itself and the return spring omitted for clarity. Note that the return spring is essential to take up backlash.

The other point is that difficulties exist in fixing the springs to the blocks B1 and B2, in such a way as to avoid high stresses or minute movements of a stick-slip kind which would cause errors and fretting. This kind of problem is considered later.

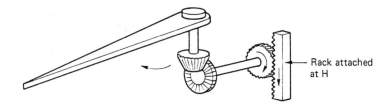

Figure 2.13   Indicator

Alternative concepts

Before leaving the bathroom scales, it is worth looking briefly at the wider possibilities. For instance, it would be possible to use an electronic display for the weight, either based upon the same mechanical parts and measuring displacement or strain, or using some different basic principle. In any case, it would be necessary to prevent errors due to the position of the load on the platform. The parallel leaf spring arrangement has this great virtue, that the central deflection is almost independent of load position: one useful way of looking at this is that it has the kinematic property that when depressed the platform moves down without tilting. It has one degree of freedom very nearly, a pure vertical translation without rotation. This kinematic property is shared by most weighing machines, but not by the traditional balance with its hanging pans.

If we were to use the strain in the springs as a measure of weight, by means of resistance strain gauges, we should have to choose a pattern of positions for those gauges which was insensitive to where the load was on the platform, and this is an interesting exercise but not a difficult one. However, once we have gone to strain measurement, other possibilities also arise. We do not need a platform which moves parallel any longer: we can, for instance, support the platform on three or four separate springs, which can be very stiff now, and add the loads in them simply by connecting strain gauges in series. The difficulty now lies in designing these stiff, strain-gauged springs, or load-cells, so that they are both accurate and cheap to make, or alternatively making them merely consistent in performance and building corrections into the electronics.

Yet another possibility is to use a system of levers to ensure that there is a single rod in the machine which carries a load proportional to the entire load on the platform, as is done in many platform weighing machines and weighbridges. We could then measure the strain in that rod, or alternatively, use a thin strip for a rod, set it vibrating and use the frequency as a measure of the load, which requires a squaring operation since the frequency will be proportional to the square root of the load.

The possibilities are many, but the point to grasp is that once we have introduced any electronics, we have multiplied the options. Frequently the best solutions will simplify the task of the mechanical parts and put the complexity into the electronics. This is the essence of much design in the 'mechatronic' field, the combination of electronic and mechanical resources in the best ways. True mechatronic designs of this sort are often very elegant and their functioning is usually very satisfying aesthetically.

2.10 SPRINGS

The energy storage aspect

The bathroom scales depend for their action on the deflection of the double leaf spring under the weight of the user. It is necessary to have enough deflection to be measured accurately, and the descending weight does work on the spring which stores it as strain energy. Thus the storage of strain energy is essential to the function of providing a measurable indication of the weight. In a similar way, all springs function essentially as energy-storing elements: the valve springs in an internal combustion engine store the energy needed to close the valve, and the spring in a vehicle must store energy so that when the wheel has risen as it passed over a bump that energy is available to move the wheel down again. Later it will be seen that a bolt is effectively a very stiff spring whose function involves storing energy.

Most springs are linear, that is, the plot of the force F in them against the deflection x, as in Figure 2.14, is a straight line, or

$$F = kx$$

where the constant k, the slope of the plot, is called the stiffness. The energy stored is the area under the plot, or

$$\tfrac{1}{2}Fx = \tfrac{1}{2}kx^2 = \frac{F^2}{2k}$$

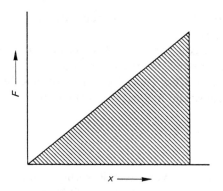

Figure 2.14 Energy stored in spring

The energy is stored as strain energy in the material of the spring. In a spring where the material is stressed in tension or compression only, as in the bathroom scales or the mainspring of a clock, the energy per unit volume is

$$\tfrac{1}{2}\sigma\epsilon = \tfrac{1}{2}E\epsilon^2 = \tfrac{1}{2}\sigma^2/E \qquad (2.28)$$

where σ is the stress, ϵ is the strain and E is Young's Modulus. The parallel between these two equations is clear.

Efficiency of use

In most springs, σ will vary through the volume of the material. In the mainspring of a clock the material is in bending, so that when fully wound σ will vary from a tension of f on one surface (the outside surface in the coil) to a compression of $-f$ on the other, and the mean value of σ^2 will be $f^2/3$. Thus the *average* energy stored per unit volume is only one-third of the *maximum* energy the material can store, or the *efficiency of use* is only one-third.

In the bathroom scales, when the diamond-shaped pieces have been cut out to make the section modulus everywhere roughly proportional to the bending moment, the maximum bending stress at all sections is roughly the same, at f, as it is in the clock spring, so the efficiency of use is one-third. If the springs had been left as rectangles, however, without the diamond shaped cut-outs, the bending stress would have been proportional to the distance from the mid-length, and the average strain energy throughout the material would have been one-third of one-third that in the top and bottom surfaces at the ends, an efficiency of use of only one-ninth. This can be seen another way: with the diamond shapes cut out, the deflection, and hence the stored energy, was 1.5 times greater, and the volume of material was reduced to one-half, so the stored energy per unit volume was increased three times.

The material in a coil spring used in tension or compression has an efficiency of use of one-half (a little less in practice, because of 'dead' coils at the ends) but stores less energy per unit volume in the most highly-stressed parts, because the material is working in shear.

A vehicle spring

Consider the mass of a vehicle coil spring using a steel of density 7800 kg/m^3 at a design stress in shear of 400 N/mm^2. Suppose the suspension has a travel from no load (with the rebound stop removed so the spring can extend fully) to fully compressed, of 0.2 m, and also that the load required to produce full compression is to be 4 kN (excluding the load carried by the bump stop). The energy to be stored is $\frac{1}{2} \times 0.2$ m $\times$ 4 kN or 400 J. The energy per unit volume in the surface, the most highly stressed part of the spring, will be

$$\frac{\tau^2}{2G}$$

where τ is the shear stress and G is the shear modulus (about 80 kN/mm^2). Evaluating this expression, the maximum energy stored per unit volume is

$$\tfrac{1}{2} \times (400 \text{ N/mm}^2)^2/80 \text{ kN/mm}^2 = 1 \text{ N mm/mm}^3$$

or 1 J per cubic centimetre. Since the efficiency of use is only one-half and the energy to be stored is 400 J, the mass of the spring must be

$$2 \times 400 \times 7.8 \text{ g} = 6.2 \text{ kg}$$

Because the end coils are not effective and for other reasons, the spring is likely to come out at about 7 kg, perhaps 1 per cent of the vehicle weight.

Matching of springs

Notice that this line of argument results only in the mass of the spring. Further progress requires a consideration of the geometry of the suspension. Figure 2.15a shows the essentials of such a suspension, of the independent trailing arm kind discussed in Chapter 1. The wheel centre is at C and the arm CAB is hinged to the body at B, while the coil spring acts at A. If the point A were moved closer to B, the stiffness of the spring would need to increase but its travel would be reduced, leaving the stored energy and the active mass unchanged. In practice, this would probably mean increasing both the coil diameter and the wire diameter, while reducing the number of turns, so putting the mass up slightly because of the greater relative importance of the end turns and a slight reduction in design stress in the thicker material.

An important point about the design in Figure 2.15a is that the point A is below CB, which not only helps with the problem of providing space for the spring but also confers some useful non-linearity. The stiffness at the wheel depends upon the moment arm h (see Figure 2.15a) of the force in the spring about B. As the wheel rises relative to the body, the arm h and hence the stiffness increase. It should be noted, however, that there is also a stiffness due to the force F in the spring and the fact that A is below B by an amount s (see Figure 2.15a), and that this stiffness, which is analogous to the stiffness of a pendulum, decreases with the rise of the wheel. It is left to the reader, if so inclined, to study the case further, but the net effect, if s is large enough, is to increase the stiffness as the wheel rises, giving a hollow characteristic as in Figure 2.15b. This is desirable, because it means that the stiffness of the springing increases with the load in the vehicle, and it also reduces the energy stored, and hence the mass of the spring.

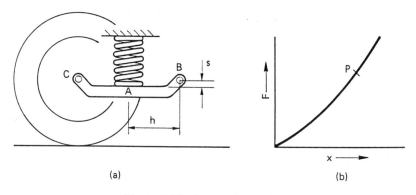

Figure 2.15 Suspension spring

It should be noted also that the spring in Figure 2.15a is not only compressed when the wheel rises, but also bent and sheared. Because the stress will then vary from turn to turn of the spring, the tendency is to reduce the efficiency of use and so increase the mass, but by making the spring straight at somewhere approaching full compression, the effect can be kept small. The alternative of providing pivots at the ends of the spring to enable it to remain straight throughout is a remedy worse than the disease. The whole of this aspect of suspension design is an interesting problem in matching (Section 1.7).

2.11 STRUCTURES, SPRINGS, ENERGY AND STIFFNESS: PERTINACITY

A useful concept for designers is that of 'holding together power', or, as it will be called here, *'pertinacity'*, the integral of force by distance throughout the structure. Maxwell showed that this quantity was a constant for a given load pattern, whatever the form of the structure carrying it.

Pertinacity is a coinage, in this use, by the writer. It is less clumsy than 'holding-together-power', the technical use is very close to the everyday use, its everyday use is rare but sufficiently common to be known, and its derivation is perfect.

As an example, consider a thin-walled pressure vessel. If we count the contents as part of the structure, there are no external loads so that the pertinacity is zero. Each cuboid element of volume of the contents, dx, dy, dz, is carrying a compressive force parallel to each of the three axes over lengths dx, dy and dz of $-pdydz$, $-pdzdx$ and $-pdxdy$ respectively, so this part of the pertinacity is

$$-3pdxdydz = -3pV$$

where V is the volume contained. Now since the overall pertinacity is zero, this implies that the pertinacity of the shell is $3pV$.

Suppose all of the shell is in two-way tension at the design stress, f. If the total volume V_S of the shell material is all in this state, then the pertinacity is

$$2fV_S$$

and to meet the requirement of zero net pertinacity

$$V_S = \frac{3pV}{2f} \qquad (2.29)$$

A spherical vessel of radius R and (small) thickness t will be stressed in uniform plane tension $(f, f, 0)$ if

$$t = pR/2f \text{ (see equation (2.22))} \qquad (2.30)$$

Also, the contained volume of such a vessel is $4\pi R^3/3$, its surface area is $4\pi R^2$ and the thickness is t, so that

$$V = 4\pi R^2 t$$

Putting the values of V and V_S into equation (2.30):

$$4\pi R^2 t = \frac{3p}{2f} 4\pi R^3/3$$

gives

$$t = \frac{pR}{2f}$$

just as in equation (2.22).

If a long thin cylindrical vessel is designed for a hoop stress of f, the longitudinal stress will be $f/2$, and the pertinacity $3f/2$ per unit volume. A similar calculation will give equation (2.20).

We can imagine the pertinacity PE divided into two parts, that representing the integral of tensile forces with respect to length through the body, PET, and that representing the integral of the compressive forces with respect to length, $-PEC$ (where PEC has been expressed as a positive quantity). Then Maxwell proved that

$$PE = PET - PEC = \text{a function of the applied load system}$$

It follows that a minimum value of $PET + PEC$ will be obtained when PEC is a minimum. It follows also that if a structure has no compressive stresses at all, that is, if PEC is zero, then PET is a fixed function of the loading system, and all structures of this kind in which all the material is fully stressed are optimum structures.

An example

Consider the loading system and structures shown in Figure 2.16. The structure at (a) is a simple tie. The pertinacity is simply the load P times the length $2L$, or

$$PE = 2PL$$

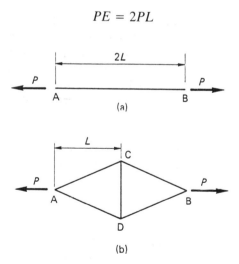

Figure 2.16 Two structures

Now in the structure at (b), the members AC, AD, BC, BD are all in tension: their length is $L \sec \theta$ and the tension in each is $(P/2) \sec \theta$, so that

$$PET = 4 \times L \sec \theta \times \frac{P}{2} \sec \theta = 2PL \sec^2 \theta$$

The single member CD is $2L \tan \theta$ long and in compression, with a compressive load $-P \tan \theta$. Since in PEC we take compressive forces (the only kind) as positive, then

$$PEC = 2L \tan \theta \times p \tan \theta = 2PL \tan^2 \theta$$

Hence

$$PE = PET - PEC = 2PL (\sec^2 \theta - \tan^2 \theta) = 2PL$$

This shows that PE is indeed constant in this case, and that if θ is increased, the increase in PET is matched by an equal increase in PEC.

If we are concerned with a material in which the design stress is f in tension and $-f$ in compression and if these limits are everywhere reached, then the volume of material used is

$$(PET + PEC)/f = (PE + 2PEC)/f \tag{2.31}$$

and since PE is invariable, reducing the mass is equivalent to reducing PEC. (Note that if $PEC > PET$, it may be tidier to look at reducing PET: it comes to the same thing.)

The energy stored in a flywheel is half the pertinacity. To see this, note that the flywheel can be designed to have only tensile stress in it, and that for any structure in tensile stress only, the pertinacity depends only on the load system applied to it. Consider a flywheel running at angular velocity ω, and a particle of mass m in it at radius r. This mass will have a centripetal acceleration $\omega^2 r$, and so will exert a centrifugal force $m\omega^2 r$ on the structure. Note that this use of the term centrifugal force is perfectly correct. Imagine this force carried by a spoke of length r to the centre, where it is balanced by other such forces. The pertinacity in the spoke is (force × distance) or $m\omega^2 r^2$. But the kinetic energy of m is

$$\tfrac{1}{2} \text{ mass} \times \text{speed}^2 = \tfrac{1}{2} m\omega^2 r^2$$

Thus this particle of mass has a kinetic energy equal to just half the pertinacity it demands in the structure to hold it in place. Since the same is true for any other particle of mass, the kinetic energy of the flywheel is just half the whole of its pertinacity.

The link with energy

In a structure which is fully stressed everywhere to the design stress f, the stored energy is, from equation (2.28):

$$f(PE + 2PEC)/2E \tag{2.32}$$

Now if the external loading consists of a single applied force F and a number of reactions at fixed points, then the work done on the structure by F as it increases from zero to F is

$$F\delta/2$$

where δ is the deflection of the point of application of F in the direction of F. Thus

$$\delta = \frac{f(PE + 2PEC)}{EF} \tag{2.33}$$

This equation shows that a more efficient structure, that is, one with a lower value of PEC, will deflect less. Thus efficient structures are also stiff.

In practice, these results will be modified by efficiencies of use (Section 2.10), but these are often uniform throughout and so alter the arguments which follow only by the addition of a constant multiplier.

For a spring, we require δ to be large and PE is fixed, so we must make PEC large. A spring can thus be regarded as an inefficient structure, in which both PET and PEC are large: in other words, there are roughly equal amounts of material in tension and compression. It is clear this is so in leaf springs or coil springs in torsion. In coil springs in tension or compression, the wire is chiefly in torsion, so the material is in shear, which is equivalent to principal stresses of $(\sigma, -\sigma$ and $0)$, and the pertinacity per unit volume is zero.

The engineering scientist may say 'doubling the material in a structure will double the stiffness', because he imagines all the scantlings being doubled, keeping the general form the same.

The engineering designer, who uses all the material at its design stress as far as is practicable, may say, 'halving the material in a structure will double the stiffness', because he imagines the form being changed to make the structure more efficient, while throughout the material is used to the full, at the design stress f.

Both viewpoints are sensible ones, but the first is characteristic of science, and the second, with its concern with purpose, of engineering.

3

Abutments and Joints

3.1 INTRODUCTION

This chapter deals with a number of important considerations in design, many of a rather detailed nature. It looks at the way parts are fitted and joined together, derives some principles about the design of mating surfaces or abutments, clarifies the function of screw threads and broaches the important subjects of register and sealing. Some of the influences which bear on the designer in these areas relate to the structural aspects considered in the previous chapter, but others derive from manufacture and assembly, or from functional requirements other than sustaining load.

The concerns of this chapter stem from the need to build designs from assemblies of parts, because a gear box, for instance, has to come apart to put the gears inside, because one material does not combine all the required properties, including low cost, and to make manufacture easier.

3.2 THE FORM OF ABUTMENTS

The influence of manufacture

Generally, the major surfaces of engineering components are flat planes or surfaces of revolution because these shapes are readily generated by machining. Even when parts are made by replicative processes like moulding, the dies or moulds are machined from the solid, so the generalisation applies to them, too. The designer should only depart from planes and surfaces of revolution with very good cause.

Of surfaces of revolution, excluding minor features such as chamfers and fillets, the cylinder is much the most common. Most engineering components are largely defined by planes and cylindrical surfaces, and it is these we chiefly need to consider in fitting *parts* together to form *assemblies*.

Bushes

Figure 3.1a is the idealisation of two abutting components, a casing 1 and a bush 2 fitting into it. Potentially, there are three pairs of fitting surfaces, two planes each at A and B, and two cylindrical surfaces at C. First of all, we should not try to make both pairs of planes at A and B fit together: this would require very close tolerances, probably individual fitting, and is almost never desirable. It is sometimes called 'making two faces', a thing to be avoided. We should decide which pair are going to fit together, probably those at A, and make a definite gap at B.

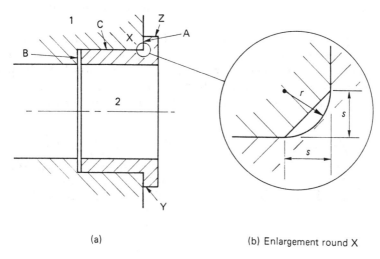

(a) (b) Enlargement round X

Figure 3.1   Abutment of bush and casing

The next point to consider is over what area we want the surfaces to touch. It is difficult and so expensive to keep a 'flat' surface flat right into the corner, and sharp corners produce high stress concentrations. We therefore provide the bush with a generous fillet radius at the corner X where the plane A and the cylindrical surface C intersect (see Figure 3.1b). It would be expensive and pointless to make the casing fit into this fillet, so we provide the casing with a chamfer which bridges the fillet.

Consider the bush at the point Y. If we left the corner sharp, it might easily be damaged and spoil the fit on the plane A. We therefore provide a chamfer at Y. Even corners like Z are chamfered, to make the parts safe and easy to handle.

Abutments

This discussion has established an important piece of good practice. All abutments should be between defined areas of the surfaces concerned, bounded by reliefs such as chamfers at which contact ends. Internal corners should have fillet radii, and mating chamfers should have a side at least equal to the fillet radius to avoid contact with the uncertain surface adjacent to the fillet, that is, in Figure 3.1b, $s > r$. These rules are an embodiment of the principle of clarity of function.

Should we limit the contacts on the surfaces A and C even further? The answer is, 'quite possibly, depending on the nature of the case'. If the bush is a small and unimportant one, the answer is probably 'no', but if it is a large and important one, further thought may be justified. Consider the three versions shown in Figure 3.2, in which the surfaces C have been relieved in various ways. In each case the bush is retained by a ring of screws at S.

In Figure 3.2a the contact on surfaces C has been restricted to a short length just beyond the flange. For most purposes this will be enough. Clamping together the plane surfaces A with screws will hold the parts together sufficiently and only a short engagement of the surfaces C is enough to locate the bush centrally, that is, to provide a register.

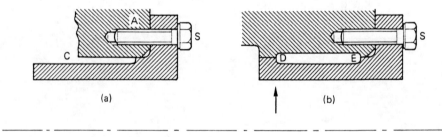

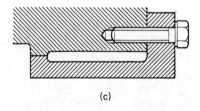

Figure 3.2   Relieved abutments

ABUTMENTS AND JOINTS

The arrangement shown in Figure 3.1b might be justified if there was a bearing at the position indicated by the arrow, subject to large transverse loads from a shaft. It is usually desirable that such a bearing should be supported as directly from the casing as possible, and so it is preferable to have a close fitting contact at D. To maintain a close fit on a long length is expensive and may lead to difficulties in assembly. Notice also that if the diameter at D is made slightly smaller than that at E, then it is not necessary to force the bush at D through a tight point at E. This raises the question of relieving the surfaces at E, too, as in Figure 3.2c, and this may be a good answer. Notice that the relief in the casing from D to E could be cast in, but would be difficult in die-casting.

It is worth thinking abstractly about this question of mounting a bush in a casing. The contact of the surfaces at A would fix the bush, apart from its angular position and the small amount of sliding of one face on another that the clearances of the screws in the holes in the flange would allow, that is, the bush is left with three degrees of freedom, one rotational and two, very limited, of translation.

Fitting anywhere in the length of the cylindrical surface C will suffice to locate the bush in translation, and very often the angular position will not be critical to within the play allowed by the screws.

Figure 3.3 shows an application of this philosophy to a bush mounted in a hole in a casting. Machining is limited to a spot face at F and a short bored length at D. This approach to manufacture is discussed again in Section 6.2.

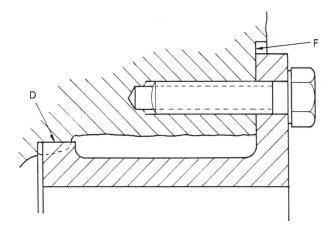

Figure 3.3 Location of bush in casting by local machining

This preliminary small study of abutments will be reinforced later by examples: the general principles involved are of constant value in design. Above all, think of problems of this kind in terms of adequate but not superfluous location, a topic which is developed more generally in the next chapter.

3.3 SCREWED FASTENINGS

Very many forms of fastening are used in engineering of which the most important groups are screws, bolts and studs. Three points about screwed fastenings need to be grasped by the designer: that they are primarily *clamping* devices, holding parts together by friction, that the important force/torque relation in them is determined mainly by friction, and that they require to store strain energy to function.

The essential function of most screw fastenings is to exert a clamping force, holding other components together with sufficient normal force to prevent relative motion. To ensure this, it is becoming more and more general to control the tightening torque which is applied to them: a short study will be made of the torque/clamping force relationship.

Torque/force relation

Figure 3.4a shows a part of the thread of a bolt which has been tightened until the tension in it is F. This force F is reacted mostly by a normal force R distributed all over the loaded face of the thread, the lower one in the figure. Consider a small patch P of this face, on which a small part δR of this force R acts. In Figure 3.4a, this force does not lie in the plane of the paper, but is inclined to it at an angle α_N, approximately 30°, which is the flank angle of the thread. Figure 3.4b is a view along the thread, in the direction of the arrow S, which shows the angle α_N which is the flank angle in a *normal* section of the thread, not in an axial section, which would be slightly different. In this view δR *is* in the plane of the paper, but in Figure 3.4a the component in the plane of the paper is just $\delta R \cos \alpha_N$, as shown.

Let μ be the coefficient of friction. If the nut has just been tightened, it has been moving to the left at P so the friction between nut and bolt will have been generating a force $\mu \delta R$ on the patch P, in the direction shown, along the threads. If the pitch angle of the threads is λ, as shown in the figure, then the axial component of δR and the friction force $\mu \delta R$ is

$$\delta R \cos \alpha_N . \cos \lambda - \mu \delta R \sin \lambda \qquad (3.1)$$

ABUTMENTS AND JOINTS

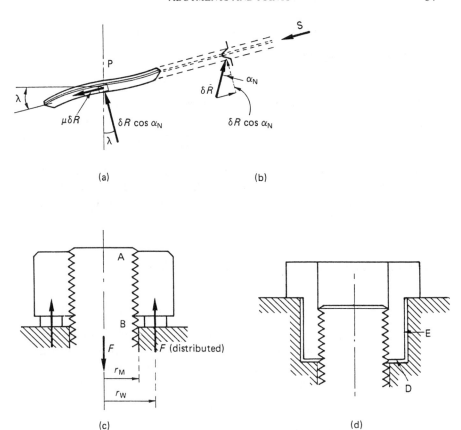

Figure 3.4 Forces and strains in screwed fasteners

and the sum of all these elementary forces along the axis must just balance F. Summing over all the δRs, since α_N and λ are constant, by axial equilibrium:

$$R(\cos \alpha_N \cos \lambda - \mu \sin \lambda) = F \tag{3.2}$$

Now consider the torque due to these forces resisting rotation of the nut. Both the δR and $\mu \delta R$ forces have a torque about the axis. They both act at radius r_M, the pitch radius of the threads. Their combined components in the tangential direction are

$$\delta R \cos \alpha_N \sin \lambda \quad \text{and} \quad \mu \delta R \cos \lambda$$

The elementary force δR also has a radial component $\delta R \sin \lambda$, but this produces no torque because it goes through the axis. The torque Q_T resisting rotation of the nut from the patch P is thus

$$(\delta R \cos \alpha_N \sin \lambda + \mu \delta R \cos \lambda) r_M$$

and this can be integrated over R as before to give

$$Q_T = R(\cos \alpha_N \sin \lambda + \mu \cos \lambda)$$

Substituting for R from equation (3.2) gives

$$Q_T = \frac{\cos \alpha_N \sin \lambda + \mu \cos \lambda}{\cos \alpha_N \cos \lambda - \mu \sin \lambda} Fr_M \qquad (3.3)$$

There is also a torque resisting rotation of the nut from the friction between it and the washer. There is a reaction F between nut and washer, acting at a mean radius r_W and producing a friction force μF and a resisting torque Q_W, say, where

$$Q_W = \mu F r_W$$

The total torque required to tighten the nut is thus

$$Q = Q_W + Q_T = Fr_M \left\{ \frac{\cos \alpha_N \sin \lambda + \mu \cos \lambda}{\cos \alpha_N \cos \lambda - \mu \sin \lambda} \right\} + \mu F r_W \qquad (3.4)$$

This equation relates the tightening torque Q to the axial load in the bolt F, and gives us the information we need to set the torque loadings to which screwed fastenings should be tightened. It is not a very transparent form, and for bolts it can be closely approximated by

$$Q \simeq Fr_M(\tan \lambda + \mu \sec \alpha_N) + \mu F r_W \qquad (3.5)$$

(usually within 1 per cent) in which each of the three terms is seen to have a simple physical significance.

Interpreting the torque/force relation

The first term is the *thread-climbing* term, which is associated with the nut advancing down the threads against the force F: it is independent of μ and α, and could be found by this argument. In the absence of friction, in one rotation Q would do work equal to the force F times the distance advanced, or

$$2\pi Q = F \times \text{pitch}$$

But the pitch is equal to $\tan \lambda$ times the circumference $2\pi r_M$, giving

$$Q = Fr_M \tan \lambda$$

The term $\mu F r_M \sec \alpha_N$ is the torque needed to overcome friction in the threads and is just like the washer friction term $\mu F r_W$ but for the factor sec α_N, which is the effect of the inclined flank of the thread. The approximate equation (3.5) is thus readily understood, and it is easily shown that in a typical fastener thread, where λ is about 3°, the friction terms dominate, giving rise to at least 80 per cent of the torque. It shows, for example, that a differential screw will not improve the mechanical advantage much over a plain screw, in spite of the very small advance per turn that can be achieved. Equation (3.5) is a much more *insightful* form than equation (3.4), and should be preferred by the designer for most purposes: it encapsulates the relationship in a particularly lucid way, and is usually quite accurate enough for practical purposes.

Designers should always try to shape their equations into such transparent forms, so that their import can be understood thoroughly.

Big end

Consider a connecting rod big end, made in two halves fastened together with two bolts (Figure 3.5a). The big end itself is a continuous circular beam subject to a variety of loads from the connecting rod, the crankshaft and its own inertia, perhaps to a pattern rather as shown. Ideally the beam would be one piece of metal, but the needs of assembly dictate that it must contain a diametral joint. The function of the bolts is purely to maintain a compressive stress in the joint faces, and to do this they should ideally have a low stiffness, that is, they should be very stretchy, so that when deformations occur due to stress or temperature that prestressing force changes as little as possible, both to maintain the bending strength of the beam at the joint and to protect the bolts themselves from fatigue. To achieve this stretchiness, the bolts are made long and are thinned down or waisted over most of their length, leaving just a short section at each end of larger diameter so as to provide a register (Figure 3.5b). Where the bolts have also to provide the register between the two halves of the big end, the centre is also left at the full diameter to act as a dowel.

It is useful to think of bolts and screws in an abstract but homely way, as very short, stiff, strong pieces of string pulling things tightly together. They should usually provide a clamping force alone: if a register is required as well, it is usually better to entrust this function to dowels or spigots, rather than using fitted bolts working in shear.

Threaded fasteners can fail in tension or by shearing of the threads. Waisting of bolts can be carried out down to the diameter at the bottom of the threads (the core diameter) without weakening them (this may not be quite true with threads made by rolling). Care needs to be taken not to make sharp transitions in diameter, which give rise to stress concentrations.

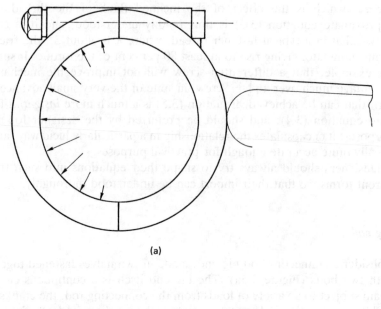

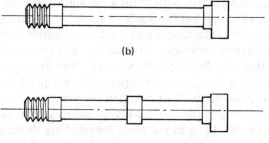

Figure 3.5 Big end of connecting rod

3.4 JOINTS

The connecting rod big end has already introduced the subject of joints. They are an unfortunate necessity, and we should avoid them where possible. If there must be joints, they should be as few as we can manage with. In the design of a joint, we should aim at making it behave as far as we can as if it was not there. Thus in the big end, we aim at a high enough load in the bolts to keep the faces of the joint solidly in contact under the worst conditions, so the ring beam of the big end behaves as if it were a continuous solid ring, without joints. To achieve this effect, it may be advantageous to relieve the mating faces in the joint. Consider a joint with rectangular faces $b \times d$, subject to a bending moment M and a direct compressive load P. A fraction k of the depth d has been relieved (see Figure 3.6). The relevant second moment of area I is

$$\frac{bd^3}{12}(1 - k^3)$$

so the maximum tensile bending stress σ_b is

$$\sigma_b = \frac{6M}{bd^2(1 - k^3)}$$

To produce an equal but opposite compressive stress over the whole area $bd(1 - k)$ requires a force P where

$$P = \frac{6M}{(1 + k + k^2)d}$$

and decreases with increasing k. For example, if $k = 0.6$ (a sensible proportion), the value of P required is reduced by 49 per cent from the unrelieved case. In a real example, like the big end joint, the saving will be slightly less, because of the relief that must be present in the form of the bolt hole.

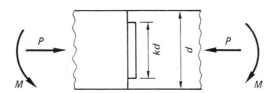

Figure 3.6   Joint with relieved abutment

It is worth staying with this example a little longer: at the joint we need to provide a sufficient bending strength in the section, a sufficient prestress to keep the section working in bending, and also a register and sufficient shear strength. Two ways of doing this are shown in Figure 3.7. Note that a good way of machining such forms of joint face is by broaching, and that consideration should be given to whether the two parts can be assembled in two ways or one, that is, whether the face is symmetrical or not.

It should be noted that the big end joint face is a disposition problem. The commodity to be shared out is the area, and the functions to be accommodated are bending strength, the prestressing force in the bolt, the register and sufficient shear strength. Here the problem is trivial: the bending area is provided where it does most good, in two 'flanges', the bolt goes in the middle, and the other two functions where they can be put, since their position does not alter their effectiveness.

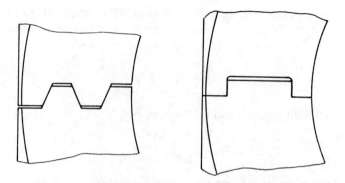

Figure 3.7 Joint face for big end

3.5 STATOR BLADE FIXING

Figure 3.8 shows a fixing for a gas turbine stator blade which has much in common with the connecting-rod joint. The blade, its platform and the screwed extension which fixes it in place in the turbine casing are cast in a superalloy. The principal load is the gas force which tends to bend the

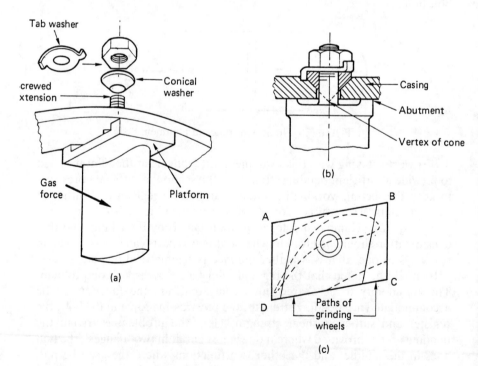

Figure 3.8 Stator blade fixing

blades in the direction shown in the figure. The joint face between the casing and the blade platform must be prestressed by tightening the nut on the screwed extension, ideally to a high enough level to keep the net stress compressive everywhere under the gas loading.

If σ_b is the maximum bending stress, Z the section modulus of the joint face, and A the area of the joint face, P the acceptable load in the extension, and the required prestress is σ_p, then if M is the bending moment caused by the gas we require

$$\sigma_b + \sigma_p < 0$$

or

$$\frac{M}{Z} < \frac{P}{A}$$

If the maximum distance from the principal axis of the joint face is y_m, and Ak^2 is the second moment of area, we have

$$\frac{My_m}{k^2} < P$$

In other words, the larger k^2 is, the better.

Looking at the view of the blade down the axis of the extension (Figure 3.8c) we see that the highest relevant k^2 would be obtained with two triangular patches adjacent to the corners A and C, as shown by the broken lines. However, we must bear in mind the manufacturing process used to finish this surface, which might be grinding or turning with relative motion in the direction shown, and thus is obstructed by the screwed extension. Also, we must not leave a leakage path for the gas from front to back of the stator blade row, so it makes sense to keep a small strip down the edges AD and BC. This leaves the best pattern somewhat as shown in the figure.

An arbitrary decision has been made and needs reviewing: the extension has been put in the centre of the parallelogram ABCD. Is this the best place? If the extension were nearer DC, it would produce more prestress at C and less at A, and it is at C where the maximum bending tensile stress occurs (σ_b). On the other hand, the extension needs to pull on the solid deep metal of the blade form, and this is nearer AB. With the other degree of freedom, the position relative to AD and BC, movement towards BC is favourable on both counts. To resolve the matter fully a finite element stress analysis is needed.

One further aspect of the stator blade is of interest. The superalloy has a much higher coefficient of thermal expansion than the material of the casing. If nothing were done, the effect of heating up would be to slacken the screwed fixing. To overcome this, a conical washer of superalloy has been provided (see Figure 3.8, a & b) fitting in a conical seat in the casing:

the vertex of the conical fitting surface lies in the cylindrical joint surface between casing and blade, and because expansion and contraction take place directly along the generators of the cone, the prestress is maintained as the casing heats up. Strictly, this requires that the nut and even the tab washer have the same coefficient of expansion as the superalloy.

The stator blade has shown several important aspects of form design and exemplified clarity of function and the avoidance of arbitrary decisions. Note also that the use of the geometric property of cones should be thought of in that way, as a useful dodge that can be used in many circumstances, not simply as a bit of turbine design practice: the abstract way of looking at design problems and solutions is the most helpful to creativity.

3.6 JOINT EFFICIENCY

The concept of joint efficiency is often useful: it is defined as:

$$\frac{\text{strength of the joint}}{\text{strength of a continuous section}}$$

Consider the screwed tension joint shown in Figure 3.9 (actually two joints back to back). Its efficiency would be 100 per cent if when we tested a number to destruction in tension, half failed in one of the smooth rods and half failed in the threads or the collar. If the average failure load was 80 per cent of that of the smooth rod, the efficiency would be 80 per cent. Notice that in this case, if the thread has been cut in the rods, so the diameter of

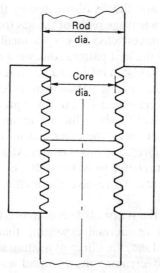

Figure 3.9 Joint efficiency

the threads is the same as that of the rods, then the efficiency is bound to be less than 100 per cent.

If the threads have been rolled, then the core diameter will not be so much smaller than the rod diameter and with the enhanced material properties developed by rolling it is possible for the joint efficiency to be one hundred per cent.

A useful concept in some joint design is the ideal scarf joint (Figure 3.10). Timber is 'scarfed' in the fashion shown, by glueing on a long inclined joint face. If we consider such a joint subjected to a tensile load T, then at section AA all the load is in the left-hand member, at section BB a little of the load is transferred to the right-hand member, at CC it is equally split, by DD it is all in the right-hand member. The tensile stress is uniform throughout.

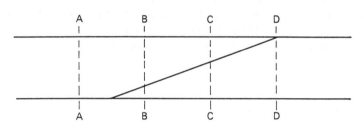

Figure 3.10 Ideal scarf joint

Now consider the screwed joint shown in Figure 3.9. Before any load can be transferred out of a rod and into the sleeve by the first engaged turn of the thread, the section has been reduced, making the joint efficiency less than 100 per cent. In order to make the joint as strong as the rod we must first increase the diameter.

In simple calculations of the stresses in the screw threads it is assumed that the load transfer by shear and bending in the 'teeth' is shared equally by all the turns. In books on machine elements it is pointed out that since the bolt is in tension along the working length of the threads and the nut is in compression (see Figure 3.4c), the pitch of the bolt thread will be increased, the pitch of the nut thread will be decreased, and so they would not fit with an evenly distributed load: hence the load cannot be evenly distributed among the turns. They go on to show that the load will be concentrated on the turns at the abutment face end, that is, at B in Figure 3.4c.

Imagine the nut being tightened. Relative to B, the bolt thread at A moves upwards because of the tension in the bolt, while the nut thread moves down, so that the loaded faces of the threads (bottom on the bolt, top on the nut) tend to separate, so that the turns at A carry less load. Analysis shows heavy concentrations of the order of 2.5 times average on the end turn at B. As we shall see, however, things are not so bad as that in most cases.

This is a problem of matching (Chapter 1). It can be overcome by the bold solution of Figure 3.4d, which throws the nut in tension too. Notice that there should be clearance at D, and probably at E too according to the principle established at the beginning of this chapter. An alternative would be to make the nut thread with a slightly larger pitch than the bolt. However, such expedients are rare, and the reason lies partly in another effect which improves the matching.

Consider the nut in Figure 3.4c. It is subject to a distributed upward force F from the washer at radius r_W and a distributed downward force F on the threads at radius r_M. Because of the difference in the radii, the nut tends to turn inside out, rolling inwards at A and outwards at B. This rolling tends to separate the threads at B and force them together at A, opposing the effects of stretch in the bolt and compression in the nut. With proportions not far from those commonly adopted, the compensation will be very good, that is, the matching will be nearly perfect.

Fir-tree root

The fir-tree root used for securing turbine blades to discs in many gas turbines (Figure 3.11) shows in a stepwise form the same graduated pattern of transfer of load as the scarf joint. The figure shows a root with two pairs of lugs. If each pair of lugs takes half the load P on the root, then the bottom neck at B will carry a load of only half P, while the top neck, at A, carries the whole load (ignoring the small loads due to the inertia of the fir-tree itself).

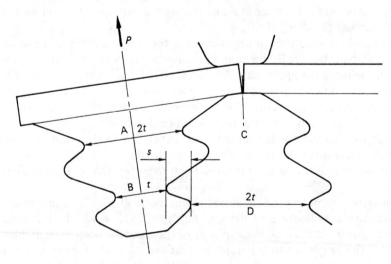

Figure 3.11 Fir-tree root

It follows then that the neck at A should be twice the width of the neck at B: let these widths be $2t$ and t respectively. Also let s be the overhang needed to accommodate a lug, and p the pitch between adjacent fir trees (see figure). If the disc and blade are of materials of equal strength and the design is balanced, then the neck C in the disc carries load $P/2$, and so should be t wide, and the neck at D should be $2t$ wide. Now into the pitch p we have to fit the width of neck A, at $2t$, a pair of overhangs of width $2s$, and the width of neck C, at t, so that

$$p = 2t + 2s + t = 3t + 2s$$

Since the entire load P has to be carried by neck A instead of the whole pitch p that would be available with, say, a perfect welded joint, then the joint efficiency η is given by

$$\frac{2t}{p} = \frac{2t}{3t + 2s}$$

Typically we find s needs to be about $0.6t$, so that

$$\eta = \frac{2t}{4.2t} = 0.48$$

If we have more lugs, say n pairs, then the widest neck will be nt wide, the narrowest t:

$$p = (n + 1)t + 2s$$

and

$$\eta = \frac{nt}{(n + 1)t + 2s}$$

or, if $s = 0.6t$:

$$\eta = \frac{n}{n + 2.2}$$

As n increases and the lugs become relatively smaller, n approaches unity more closely, as one might expect. The amount that n is less than unity is roughly

$$\frac{2.2}{n + 2.2}$$

which illustrates a common finding in design, that when we try to improve performance by increasing the number n of steps or stages the shortfall below the ideal is roughly inversely proportional to n. It illustrates also that the gains diminish rapidly with n. The same is true of the number of boiler feedwater heaters in a power station or the number of gears a car has.

68 FORM, STRUCTURE AND MECHANISM

The reader may have noticed that the fir-tree root design is a disposition problem, in which the commodity is the pitch p which has to accommodate blade and disc necks and overhangs. Efficiencies of higher than about 30 per cent require the use of overlapping, which happens when $n > 1$.

3.7 OFFSET BOLTED JOINTS: LUGS

In the big end joint, the bolt ran nearly through the middle of the section at the joint. Very often this is not possible, as in a joint between flanged pipes (Figure 3.12a).

If the width of the flange is small compared with the diameter of the pipe and the pipe itself is thin-walled, the problem becomes effectively a two-dimensional one. The stresses will approximate to those in two rods joined by bolted lugs on one side only, as in Figure 3.12b. If we imagine one bolt pitch p of the pipe wall and flange straightened out to give this figure, the effects of changing the number of bolts, for instance, can be studied in a useful way. Also many casing joints are more closely of this form, with flat walls joined by bolted flanges, exactly as in Figure 3.12b.

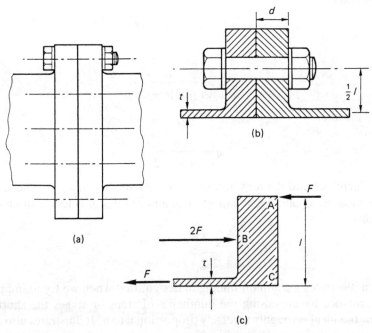

Figure 3.12 Pipe joint

We have then two rods or strips, t thick and p wide, joined by one bolt through two lugs, d thick, p wide, and standing off a length l from the strips (see Figure 3.12b). Suppose the joint is subject to a force F, and the bolt is half way along the lugs, at B (Figure 3.12c). Then the lugs will tend to separate at C, loading the bolt in tension and forcing the lugs hard together at A. We can regard a lug as a beam, loaded at A, B and C, with a maximum bending moment at B of $Fl/2$. If we neglect for the moment the bolt hole, the maximum bending stress will be

$$\sigma_b = \frac{Fl}{2} \bigg/ \frac{pd^2}{6} = 3Fl/pd^2$$

Now in a well balanced design, this will be equal to the stress σ_w in the wall and both will be equal to the design stress f, so that

$$\sigma_w = \frac{F}{pt} = f = \sigma_b = \frac{3Fl}{pd^2}$$

or

$$d = \sqrt{(3lt)} \qquad (3.6)$$

Suppose that t has been determined, and we wish to design a flange to suit. The bolt size will need to be adequate to carry the load F, and the allowable stress in it will generally be much higher than f: this will be conveniently represented by taking the effective area of the bolt cross-section as c^2, where c is the bolt diameter. We find then that

$$F = c^2 f = ptf \qquad (3.7)$$

which gives a value of p for a given c (since only standard bolt diameters are available). We must not put the bolts too close together, however, and a rough rule is that $p = 5c$ gives adequate spanner clearance. Putting $p = 5c$ in equation (3.7):

$$c^2 = 5ct \quad \text{or} \quad c = 5t$$

Now turn to l. To give spanner clearance, l should not be less than $3c + t$, or $16t$.

Modifying equation (3.6) for d to allow for the bolt hole (width c in a pitch of $5c$) we find

$$d = \sqrt{\left(\frac{3lt}{0.8}\right)} \simeq 8t$$

Figure 3.13 shows these proportions sketched as for a straight wall. Note how thick and heavy the flange appears: this is because the flange is acting as a beam, which is an inefficient structure, whereas the wall has been assumed to be fully stressed and in pure tension, and so very efficient.

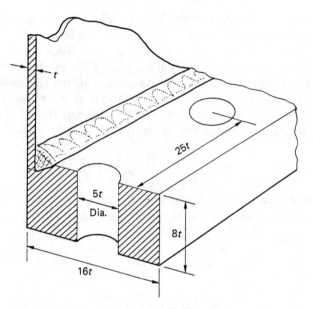

Figure 3.13 Proportions of flange

When the flange width is a substantial fraction of the diameter of a pipe, other effects become important. The pipe wall provides a helpful moment at the inner radius of the flange: the flange itself develops circumferential stresses which also help.

It is worth looking at what happens if we use a larger bolt, say, k times the diameter, where $k > 1$. The number of bolts can be reduced by a factor $1/k^2$, since each bolt is k^2 times as strong. To provide spanner clearance, l must increase by a factor k and so the bending moment in the flange goes up k times, requiring d to be increased to $k^{1/2}d$. The amount of metal in the flange goes up $k^{3/2}$ times: on the other hand, fewer larger bolts are cheaper and quicker to assemble.

An alternative is to use a number of separate lugs instead of one continuous lug or flange. These separate lugs can be deeper than the flange and yet use less material. For instance, if the width of a lug is only one-quarter of a pitch, it need only be twice the depth of the original flange to have the same bending strength, thus saving half the material (see Figure 3.14). In practice, the gain is not so great, because the bolt hole takes off more from the strength of the lug than it does from the strength of the flange.

Sometimes the extra width provided by the flange is desirable or necessary for sealing purposes. Then a common solution is to keep a thin flange, extending between deep lugs (Figure 3.15).

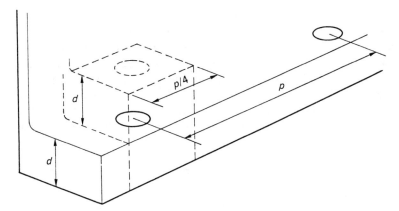

Figure 3.14   Replacing a flange with separate lugs

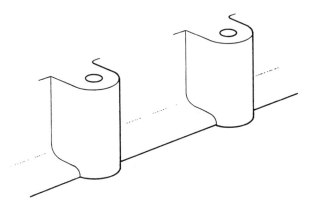

Figure 3.15   Thin flange with lugs

3.8   GENERAL PRINCIPLES APPLIED TO JOINT DESIGN

The offset bolted joint has demonstrated some general principles of structural design. One of the chief is that the designer should have a simple but clear vision of how the structure will work, of how the load will be transferred from point to point, usually based on an imagined decomposition into rods and beams. Eventually proper stress analysis is usually needed and changes may result, but design by repeated stress analysis is slow and expensive (though becoming less so), particularly if the designer's first guess is a poor one. Moreover, if the changes are large, they may have effects beyond the strength aspect, requiring more radical and expensive alterations and the waste of other work.

As a sorry example, consider Figure 3.16a, which shows the form originally used for a large rotor. Because of the need to keep the mass down and the stiffness up to avoid critical speed problems, the rotor was to be hollow and thin-walled. Because of the large diameter and the small access diameter at the ends, a joint was to be provided at J. Unfortunately, it was not realised that the radial web at A was very flexible in the mode shown in Figure 3.16b, sometimes known as 'oil-canning'. This unrecognised bending flexibility invalidated the critical speed calculations. Figure 3.16c shows a variation which overcomes the problem. A cone, even rather a flat cone, is many times stiffer in bending than a flat plate.

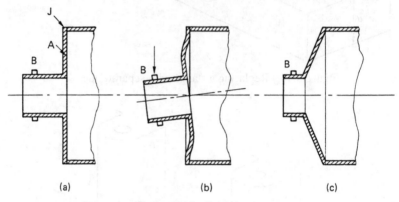

Figure 3.16 Joint in rotor

This weakness in bending was probably overlooked more easily for the following reason. A bending flexibility near the end of a shaft is much less serious than one near the middle. An illustration of this is the fact that ordinary fit human beings can support their own weight when lying horizontally with supports under their heels and the backs of their heads only (Figure 3.17). The reason is that the bending moment at the neck, the weakest point of this human beam, is quite small, and where there is the maximum bending moment the cross-section is relatively generous (too generous in some of us).

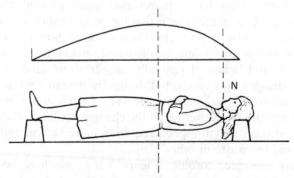

Figure 3.17 Human beam

A special consideration with a joint in a high-speed rotor is the importance the function of providing a register assumes. It has to be very precise, and very repeatable if it is ever to be demounted and reassembled. A simple spigot will not generally do: dowels will, if interchangeability is not demanded (and in many cases interchangeability may be too expensive and unimportant as well).

3.9 REGISTER IN JOINTS: INTERSECTION PROBLEMS

The problem with a joint like that in Figure 3.16c is that it governs the alignment of the two journals which in turn determines the axis of rotation. To ensure accuracy it is desirable to machine the journals with the joint assembled. If subsequently the joint must be broken and remade, it must go back very nearly exactly as it was before.

To achieve this repeatability, it is desirable that the mating planes should be very true and flat, and that is best achieved by a process in which there is an unobstructed cut over the entire surface. This is provided by the dowelled joint shown in Figure 3.18a, but not by the spigoted one in Figure 3.18b. In the spigoted joint, the surface A on the left-hand part is

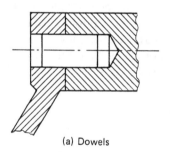

(a) Dowels

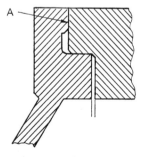

(b) Spigot

Figure 3.18 Means of obtaining a register

obstructed by the spigot, but nevertheless high accuracy must be obtained: note the desirability for this purpose of a small relief at the inside diameter as shown.

There is often an objection to a spigot on a large diameter and this illustrates a general principle of design, which might be called 'local closure'; it is closely related to the principle of short direct force paths which will be treated later. The dowel ensures that points close to the dowel on each of the parts joined are assembled accurately together. No large distances are involved with possible expansions and shrinkages that could affect the register. A spigot lacks this important virtue.

It is expensive to make dowelled parts interchangeable, so that commonly a pair of components which are dowelled together must be treated as one item thereafter. The holes for the dowels are often reamed in line with the parts assembled, and this removes the need for very great accuracy in positioning them.

There are ways of making joints of this kind which are interchangeable, by using some form of *face coupling*. Figure 3.19 shows a Hirth coupling. Each of the serrations is cut by a V-shaped cutter whose point (which is imaginary, the tip being radiused) travels on a line not quite parallel to the plane of the joint, but making a small angle α with it. The mating part is made in the same fashion. The effect is that, like the conical washer, the two parts fit each other even if one expands. Another coupling with similar properties is the Gleason Curvic Coupling. Both can be made to very high accuracy using special purpose machines.

One advantage of dowels is that the surfaces they register can be machined all over and are not obstructed. Many difficulties arise in design

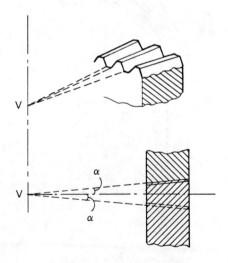

Figure 3.19 Hirth coupling

because one piece of space is needed for two purposes. For instance, the screwed extension of the stator blade intruded into the space which ideally was needed for machining the cylindrical seating surface all over, and this consideration modified the design: a slightly different form for the seating areas would have been chosen otherwise.

Difficulties of this kind are referred to as 'intersection problems' because they usually occur at the intersection of two features having different functions (Section 4.7). They often occur in connection with sealing, which is the subject of Section 3.11.

3.10 ALTERNATING LOADS ON BOLTED JOINTS

An important insight into joint design results from considering the effects of alternating loads. Towards the top of the stroke, a piston and connecting rod in an engine are decelerating, and the braking force which does this comes from the crank pin via the big end, the joints of which are thus thrown in tension. The joint between the cylinder head and the cylinder block is likewise thrown in tension opposite each cylinder every time it fires. What is the effect of these intermittent loads?

Before such an external load is applied to a bolted joint, the bolts are in tension and the other parts in compression, the loads being equal and opposite, say, F in the bolts and $-F$ in the rest. The added load G will give rise to tension loads in both, say, G_B in the bolts and G_R in the rest, where

$$G_B + G_R = G$$

The important question is, how is the load G shared? If we assume the joint does not actually open, in other words, if the combined load $-F + G_R$ in the structural parts remains compressive (or negative), then the way the load is shared will depend on the relative *stiffness* of the bolts and the rest. Figure 3.20 shows one-half of a big end joint subjected to a load G, under

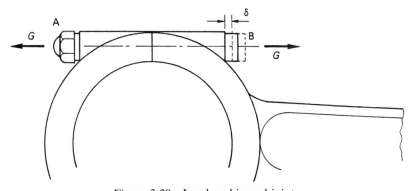

Figure 3.20 Load on big end joint

which it extends by an amount δ between A and B. If we regard the bolt as a very stiff spring, with stiffness k_B, then the change in load in it, G_B, will be

$$G_B = k_B \delta$$

Similarly, the change in load in the haunches of the big end joint between A and B will be

$$G_R = k_R \delta$$

where k_R is the (very high) stiffness of those haunches. Eliminating between these two results gives

$$G_B = \frac{k_B}{k_R} G_R$$

The bolt and the haunches share the load in proportion to their stiffnesses k_B and k_R. In general, k_B will be much smaller than k_R because both structures will have much the same elastic modulus (both are likely to be of steel), the cross-sectional area of the haunches will be much greater, and also the bolt will have additional flexibility in the ends, the nut and the head. But from the design viewpoint, we should consider what is the desirable way in which the external load G should be divided.

The answer is that we should like *all* the load to go through the haunches, reducing the compressive stress imposed by the bolt in them, rather than increasing the tensile stress in the bolt. It is the bolt which is at risk of *fatigue*, not the haunches, because of the high steady tensile stress in it and its lower cross-section, even though the fatigue limit of the bolt material may be higher. The designer will therefore seek to decrease the ratio k_B/k_R.

There is little virtue in trying to increase k_R, say, by increasing the scantlings of the haunches, because the extra mass would add to the inertia loads, the increased local *bending* stiffness would be undesirable (see Section 10.1), the outside dimensions might force an overall increase in engine length and width, and so on. The answer is to make the bolt stiffness k_B as low as possible, and this matter has already been addressed earlier in this chapter (Sections 3.3 and 3.6). This is an example of the principle of clarity of function: we aim at making the function of the bolt one of pure clamping, and to do this we make it as stretchy as possible, in pursuance of the simple philosophy of Section 3.1, now justified by the analysis above.

One aspect of importance is that in much modern practice, bolts are tightened until parts of the material yield, and it becomes desirable to design bolts able to resist repeated slight yielding without failure. This is achieved by providing zones in which high local stresses can cause yield, while redistributing the stress so as to stop the yielding process. Such zones occur at changes of section, which result in stress concentrations: some new designs of bolt have multiple circumferential grooves so they can survive repeated overstretching – that is, stretching till local yield occurs. Some readers may be puzzled by this deliberate introduction of multiple stress concentrations into a part prone to fatigue, but it must be remembered that there are always stress concentrations at the threads and under the bolt head, and the additional weak points should all be equally weak. A large number of self-reinforcing weak points produces a bolt with high robustness under repeated overstrain.

3.11 SEALING

It is frequently necessary to confine fluids to certain spaces or to keep them out of others. This problem is not too difficult when the boundaries of the spaces it is required to seal are not penetrated by moving parts such as rotating shafts or oscillating piston rods. Seals between parts which do not move relative to one another will be called 'static': where there is movement they will be called 'running' seals. A third type of seal is represented by valves and their seals, which are static when working but open in operation, and these will be called 'closure' seals. Finally, there are seals which permit movement but provide a continuous solid barrier, such as diaphragms or roll-socks, and these will be called 'hermetic'.

In the case of static seals it is important to distinguish between compliant and structural kinds. Figure 3.21a shows in section an O-ring in its groove.

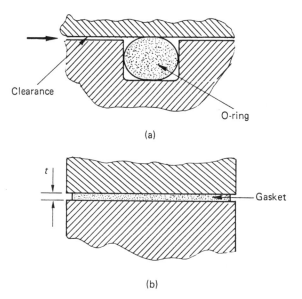

Figure 3.21 Section of O-ring and gasket

The depth of the groove is less than the diameter of the ring, which is therefore squashed into firm contact with the bottom of the groove and the closing surface. Moreover, when a pressure difference is applied to the seal, from the left in the figure, it is forced into the opposite end of the groove, improving the sealing action. At very high pressures, however, the ring may be forced or extruded into the clearance between the structural parts if this is too large.

The presence of an O-ring makes no difference to the structure of the system, unlike a gasket (Figure 3.21b). The O-ring is 'compliant', it fits into the space provided for it, whereas a gasket holds apart the faces it seals, and has a structural role: the distance t in the figure is the squashed thickness of the gasket. An important case of O-ring use is discussed in Chapter 10, the *Challenger* space shuttle disaster.

4

Freedom and Constraints: Bearings

4.1 DEGREES OF FREEDOM

A point has three degrees of freedom in space: to define where it is, we need three coordinates x, y and z, or r, θ and z. It can make *independent* movements in three directions mutually at right angles. If we connect it by a link of fixed length to a fixed point with ball joints at each end, then we remove one of its degrees of freedom. If, for example, the fixed point is the origin and the link is of length L, then the point must lie somewhere on the surface of the sphere which is of radius L and has its centre at the origin, and it has only two degrees of freedom which can be expressed as the latitude and longitude on that sphere. The co-ordinates x, y, z must always represent a point on that sphere, which will be the case if

$$x^2 + y^2 + z^2 = L^2 \qquad (4.1)$$

One equation relating the co-ordinates is equivalent to the loss of one degree of freedom.

If a point is linked by two such links to two fixed points, then there are two equations and the point retains only one degree of freedom. Finally, a third link will usually remove all the degrees of freedom and *fix* the point.

Suppose two points A and B to be a fixed distance L apart. Then between them they would have six degrees of freedom, x_A, y_A, z_A, x_B, y_B, z_B, less one because

$$(x_A - x_B)^2 + (y_A - y_B)^2 + (z_A - z_B)^2 = L^2 \qquad (4.2)$$

Thus a line AB has five degrees of freedom in space.

Consider a triangle ABC. Its three vertices have three degrees of freedom each, less three because the distances AB, BC and CA are fixed.

Thus the triangle, which may be seen as a simple body, has

$$(3 \times 3) - 3 = 6 \text{ degrees of freedom}$$

We may reach this result also by this argument: any chosen point A in a rigid body has co-ordinates x_A, y_A and z_A, to which we may freely assign values. We can also make rotations of the body about axes parallel to the three axes Ox, Oy, Oz, and these are independent motions, giving a total of six degrees of freedom.

Constraints

It is important to understand the common constraints which are used in engineering to join components into assemblies. The ball-joint ended link, rarely met with in that form but often approximated to, removes one degree of freedom. If a point is constrained to lie in a plane, then its degrees of freedom are reduced from three to two, so again we have a single constraint. If a point is forced to lie on a line, its degrees of freedom are reduced from three to one, and we have a double constraint, removing two degrees of freedom. If a point of a body is fixed, then all the three degrees of freedom of that point have been removed.

Look now at Figure 4.1a, which shows two views of a body A mounted on a platform B by means of a ball-joint at C and two links DE, FG. It may give the impression that the body is rigidly fixed, but this is not so. If we count the constraints there are five, three at C which is a fixed point, and one at each of the links DE and FG, from which it can be deduced that the body has one degree of freedom still. This is actually rotation about the axis CI, where I is the meeting point of DE and GF produced.

Figure 4.1b shows how it is possible to add a sixth constraint without achieving fixity: it shows a further link HK in the plane of the paper such that KH produced goes through I. This extra link does nothing to prevent rotation about CI. What it does do is fix the distance CI, which is already fixed by the body. If now the body becomes hot and tends to lengthen, this is opposed by the mounting: the mounting is *redundant*, but it is still not fixed.

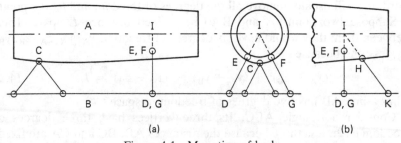

Figure 4.1 Mounting of body

The rule about constraints on a single rigid body is:

6 − number of constraints = number of freedoms − number of
redundancies   (4.3)

Very often, we want to fix the body (number of freedoms = 0) without introducing any redundancy, which might cause damage or prevent interchangeability. The designer aims in such cases at no freedoms and no redundancies.

4.2 SHAFTS AND BEARINGS

Consider a deep groove ball bearing. If the outer race is fixed in a stationary casing and the inner race on a shaft, how many degrees of freedom are left to the shaft? If very small freedoms are excluded, then the answer is one, a freedom of rotation, but the practical answer is three, because there are small freedoms of slope of the axis of the shaft which are very important.

It is common to mount a shaft in a ball bearing at one end and a roller bearing at the other (Figure 4.2a). If the shaft is straight and the bearings in good alignment, the assembly runs easily. The bearings each constrain

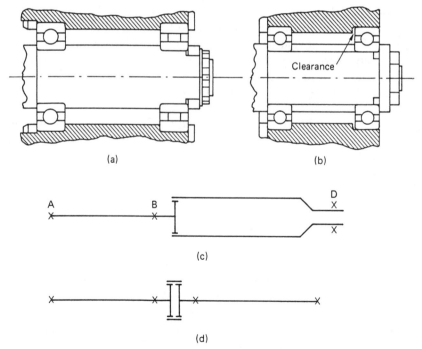

Figure 4.2 Some bearing arrangements

their journals not to move at right angles to the axis of the shaft and the ball bearing also locates the shaft axially, and is for that reason often called 'the location bearing' in such an assembly. The roller bearing removes two degrees of freedom, the ball bearing removes three, and the shaft is left with just one, that of rotation about its axis. There is no redundancy.

Sometimes a ball bearing is used at both ends of a shaft, and this arrangement can be redundant. To prevent this, the outer race of one bearing is often not fixed in the housing, but can slide a millimetre or so axially, and does not fight with the other bearing (Figure 4.2b).

Sometimes a shaft needs to be supported at more than two points. If it is very flexible, it may be enough to fit, say, two journal bearings and one location bearing (for example, two roller and one ball bearing), any misalignment of the three being taken care of by bending of the shaft. This may be checked by calculating the forces required to bend the shaft by the maximum amount the tolerances will allow, and seeing whether those forces are within the capacity of the bearings to carry, on top of any other loading.

Multiple shafts

An alternative is to break the shaft into a number of separate shafts, each with two bearings, and connect them with couplings. In general, these couplings must be able to accommodate both angular and lateral misalignment and axial movement. This arrangement is common with prime mover/load combinations, for example, engine and generator, motor and pump. Where torques are very large and extreme reliability is needed, coupling design may present serious problems, as in power station steam turbine-generator sets.

Another possibility is shown in Figure 4.2d. Here two shafts are supported on three bearings at A, B and D and linked by a coupling at C. If A and D, say, are location bearings, then A, B and D provide 3, 2 and 3 constraints respectively, and the two shafts must have their combined 12 degrees of fredom reduced to one, that of rotation. Thus the coupling C must impose

$$12 - 8 - 1 = 3 \text{ constraints}$$

These are, that it must prevent transverse relative motion of the two shafts (two constraints) and relative rotation about the common axis (one constraint). Relative axial motion must not be constrained (and this presents difficulties). An alternative is to make only one of bearings A, B and D a location bearing and to prevent relative axial movement at C.

The difficulty about couplings which allow axial movement is that when subject to high torque, the friction in them is likely to prevent any axial sliding. This difficulty may be avoided by using rolling elements or flexing elements. A case of this sort is discussed in Section 9.7.

Sometimes axial redundancies in shafting, far from being undesirable, are deliberately provided: this happens when bearings are preloaded against one another, and the classical case is that of taper roller bearings (Figure 4.3). In assemblies like that in the figure, the nut may be tightened to a specific torque, thus loading the two bearings against one another. Such installations are usually short and not subject to large temperature changes or differences, so that the locked-up load never rises to a damaging level. Note in passing that the effective centres of support of the shafts are at X and Y, that is, these are the points which should be taken in calculating bending moments and stresses. This displacement of the support to one side is valuable in extending the effective distance between the bearings, in wheel hubs or bevel gear mountings for instance. Angular contact ball bearings are also commonly preloaded, as in bicycle wheel hubs.

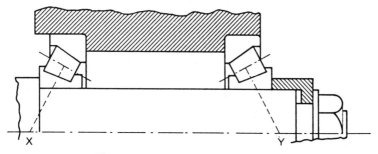

Figure 4.3 Taper roller bearings

4.3 THE PRINCIPLE OF LEAST CONSTRAINT: KINEMATIC DESIGN

A very fertile idea in design has been 'the principle of least constraint', which may be stated formally as follows:

> in fixing or guiding one body relative to another, use the minimum of constraints.

This principle was known to Kelvin, although it does not appear to have originated with him. He used it in the design of instruments, which is the field in which perhaps it first became widely used. It is used in the system of machine tool testing of Schlesinger, and had a strong influence on the early designers of gas turbines, who called design according to its precept

84   FORM, STRUCTURE AND MECHANISM

'kinematic design' (not to be confused with the classical kinematic design of linkages etc.). It must not be used blindly and invariably, as the case of preloaded bearings shows, but it is perhaps the single most useful principle in machine design.

4.4 EPICYCLIC GEARS

As an example of a more difficult kind than the shaft systems, consider the simple epicyclic of Figure 4.4, which may be viewed in the first place as a plane mechanism, in which each object has initially three degrees of freedom, two of translation and one of rotation, x, y and θ, say. The gear casing can be regarded either as a further object, or, as will be done here, a fixed frame of reference. Then we have six objects, three planets, the spider which holds them, and the sun pinion and the annulus gear which each mesh with all the planets. Now a meshing removes one degree of freedom, because one gear in contact with another can both slide and roll relative to it, or, if you prefer, because the contact between them requires only one mathematical equation to represent it.

The planets will all turn in bearings in the spider, and this removes their freedoms of translation, two each which makes six. Initially our six objects will have had three degrees of freedom each: the six meshes remove six and the planet bearings remove a further six, leaving six. If we mount the spider in bearings in the casing, that removes two more, leaving four. Now, what might be called a 'non-kinematic' approach would fix the annulus rigidly to the casing, removing a further three, and put the sun pinion in bearings, removing two more and leaving minus one. However, we can see that the mechanism is a conventional epicyclic reduction gear with one degree of

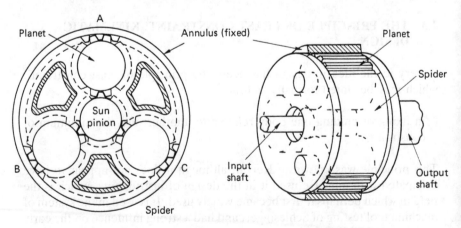

Figure 4.4 Epicyclic gear; section, and outside view with part of annulus cut away

FREEDOM AND CONSTRAINTS: BEARINGS

freedom. The sun pinion can be rotated and the spider and planets will rotate more slowly in the same direction. Thus

$$-1 = \text{number of degrees of freedom} - \text{number of redundancies} \tag{4.4}$$

so that there must be two redundancies. And indeed there are, for any two planets could be removed and the gear would still function in the same way, although the torque it could carry would be less.

Balancing freedoms

The kinematic design approach is to add only such degrees of freedom as are necessary – 'use the minimum of constraints'. With the planets mounted in bearings in the spider, the gears all meshing and the spider in bearings in the casing, there were four degrees of freedom left: since there is to be one degree of freedom left, three more constraints are the minimum needed. If we prevent the annulus from rotating, but leave it free to translate, and put the sun pinion in bearings, that will make three constraints, and we shall have a 'kinematic' design, with one degree of freedom and no redundancies. The advantage is that the tooth loads on the three planets will now be balanced.

To see this, consider the equilibrium of the annulus. Its mounting constrains it in rotation only, and so applies a pure torque Q to it (Figure 4.5a). In addition, it is subjected to three tooth forces T_1, T_2, T_3. Since they are at 120° to one another and must form a closed polygon to be in equilibrium, they must all be equal (Figure 4.5b). If the annulus gear were constrained in translation, the constraints would exert reactions R, S on it and T_1, T_2 and T_3 would not generally be equal. The distribution of load between the three planets would depend on the accuracy of manufacture and the flexibility of the components.

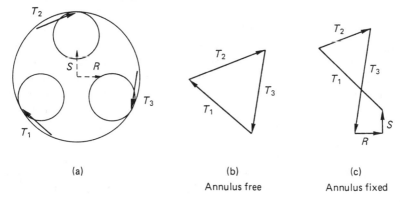

Figure 4.5   Equilibrium of annulus

Leaving the annulus free laterally means that it moves to balance the loads on the three planets. The behaviour is analogous to that of the support system shown in Figure 4.6. The three-legged spiders A, B and C stand on short pylons fixed to the base, and the object to be supported rests on the nine legs with its centre of gravity over the centre of the triangle ABC. By symmetry, the loads at A, B and C are equal, and by the same principle, so are those at each of the nine points where the object is supported. By inserting a second tier of nine smaller spiders above the first three, the loads may be distributed equally over 27 points, and so on. This method has been used for surface plates and also for the mirror of a telescope.

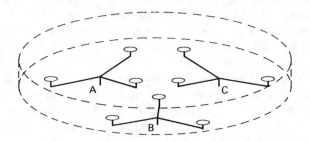

Figure 4.6 Support system

4.5 BEARINGS

It is appropriate at this point to consider various aspects of moving joints (as against the fixed joints of Chapter 3), and particularly the properties and choice of bearings. This is a large subject with its own specialised literature, and most mechanical engineering courses deal at length with the engineering science of bearings. Here the approach will be to review the 'repertoire', the range of types the designer has at his disposal.

For all but low speeds and low bearing pressures, the rubbing of one solid part directly on another leads to unacceptable wear or seizure, to say nothing of friction losses. The chief expedients adopted to avoid these problems are rolling element bearings and fluid film bearings. For limited amplitude of motion, bearings may sometimes be replaced by flexing elements, and magnetic suspension is also a possibility. Fluid film bearings can be hydrostatic, where the pressure to maintain the film is provided by a pump, or hydrodynamic, where viscous forces maintain the film pressure, either by 'wedge' action or by 'squeeze' action. Some bearings are hybrids between hydrodynamic and hydrostatic, and some hydrodynamic bearings use both wedge and squeeze action. The family of bearings is illustrated in Figure 4.7.

FREEDOM AND CONSTRAINTS: BEARINGS   87

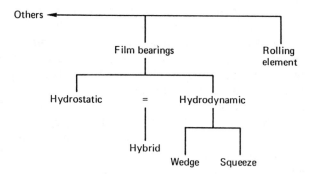

Figure 4.7   Family of bearings (part only)

4.6   HYDROSTATIC BEARINGS

The hydrostatic bearing is the easiest to understand and Figure 4.8 shows a circular pad bearing which will serve as a simple example. B is a smooth surface on which the circular pad A is to move easily. A circular recess in A is fed with high-pressure oil via a restrictor, which is simply a small short hole. The oil forces its way out through a small clearance between A and

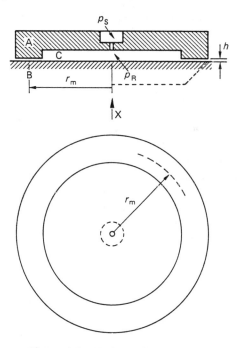

Figure 4.8   Hydrostatic bearing

B, the gauge pressure dropping in the process from p_R in the recess to zero at the outside, rather as illustrated by the broken line in the figure. This pressure forces A and B apart and maintains the fluid film in the clearance h. The net effect is the same as if the whole pressure in the recess acted over some intermediate radius r_m about half-way between the radius of the recess and the outside radius of the pad.

The function of the restrictor is to provide some stability to the system of which the pad forms part. The oil is supplied at a gauge pressure p_S, which falls to zero in two stages, to p_R as it passes through the restrictor and to zero as it leaks through the clearance h. If h is reduced, that is, the pad moves closer to the surface B, the flow through the gap is reduced and, since this is also the flow through the restrictor, the pressure drop $(p_S - p_R)$ through it will be less: p_R will rise and so will the upward force on the pad. Thus the effect of the restrictor is to make a reduction in h increase the upward force, that is, to make the bearing stiff. As we squash the pad down further on the surface, it pushes back harder.

To see why this stiffness is desirable, imagine an instrument floating on a surface on three such pads, spaced out at the corners of an equilateral triangle, with the centre of gravity in the centre. Suppose a further vertical load to be placed on the system over one pad. Then if p_R were equal to p_S, that pad would be forced down on to the surface. Even to balance the system vertically would require p_S to be precisely adjusted (there would always be a little stiffness, because of pressure drops in the oil passageways).

Power consumption

Hydrostatic bearings are lossy: they use up power in pumping the oil, and adding the restrictors increases the pressure p_S which is required and hence the pumping losses. Decreasing the clearance h reduces the flow and hence the power required, but change in this direction is limited by the truth and smoothness of the surfaces. At a given h, the flow of oil in the gap is proportional to the radial pressure drop per unit length of gap. Consider all the dimensions of the pad doubled, and the load on it and h kept the same. The area of the bearing has been quadrupled, so p_R can fall to a quarter. The length of gap is doubled, and the pressure drop along it is one-quarter, so the flow per unit circumference is down to one-eighth. As the circumference has been doubled, overall the flow in the larger bearing will be one-quarter that in the smaller one, and p_R and hence its supply pressure will be one-quarter that in the smaller one, so the pumping power, which is pressure times flow, will be one-sixteenth!

This argument suggests that we should always use the lowest oil pressure and the largest area possible, but there are other matters to be considered and we have neglected the variation of h with size of the pad. Depending on the waviness of the surface B, the practical value of h we can go down to depends on the size of the pad, smaller pads allowing smaller values of h, and as the flow rate and hence the pumping power is proportional to h^3, this effect strongly favours higher pressures and smaller pads.

We must also consider the losses due to the friction in the bearing, which increase with increasing diameter.

Another important aspect is that there may be an oil pump for other purposes, and clearly it is desirable to use one pump for all needs, if this does not compromise too much any of the functions for which the pressure is required. As a rule, the best pressure will be a middling one, say, in the range 0.3–3 N/mm².

Configuration

Most bearings, of course, are journal bearings, a cylindrical shaft running in a hole. Hydrostatic journal bearings are made by grouping round the circumference a number of recesses, each with its own restrictor and supplied with oil by a common gallery or main, as in Figure 4.9. If the shaft moves to the right, say, the pressure in the left-hand recess will decrease because of the increased flow and the pressure in the right-hand recess will increase because of the decreased flow: both effects produce unbalanced forces to the left which will try to push the shaft back into the middle.

Notice the characteristic *disposition* problem presented to the designer by a hydrostatic bearing: part of the available gauge pressure, p_S, is used to support the load, while the rest, $(p_S - p_R)$, is used to provide stiffness. The designer chooses how to divide p_S to the best effect. Notice that since the

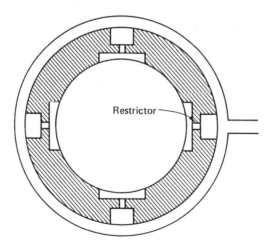

Figure 4.9 Hydrostatic journal bearing

function of the ($p_S - p_R$) part is to vary with flow, it is worth considering what form of restrictor will have the sharpest change of pressure drop with flow.

Choice of restrictor type

The clearance gap in the bearing is itself a kind of restrictor, one in which the flow is *laminar*. All the fluid is flowing in the same direction, straight down the passage (Figure 4.10), with zero velocity at the wall and a maximum in the centre, under the influence of the pressure drop and viscous stresses, with effectively no dynamic stresses, that is, no forces due to the acceleration of mass. The *velocity profile*, the variation of velocity with distance from the walls, is parabolic, as shown in the figure. The curvature of this profile (strictly, its second derivative) is proportional to the rate of pressure drop from left to right. This pressure gradient is proportional to the flow rate, as is the pressure drop in any long, thin passage where viscous forces predominate. There is a dimensionless product of parameters of the flow, called Reynold's number (*Re*) which tells us when viscous forces predominate in this way:

$$Re = \frac{\rho V h}{\mu} \qquad (4.5)$$

where ρ is the density of the fluid, V is its mean velocity, h is some typical dimension (sometimes called the hydraulic diameter, here equal to the clearance, and usually denoted by D) and μ is the dynamic viscosity of the fluid. If *Re* is less than 2000, then the flow will be dominated by viscous stresses and the pressure drop will be proportional to V. Normally *Re* will be very much less than 2000 in a bearing clearance. In flow round a ship or an aircraft *Re* may be hundreds of millions, and the flow is dominated by dynamic stresses: *Re* may be regarded as the ratio of a 'typical' dynamic stress, ρV^2, to a 'typical' viscous stress, $\mu V/h$, though 'typical' is very loosely used here.

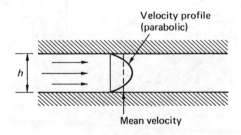

Figure 4.10 Flow in a small clearance

In the restrictor, which has the form of a small short hole and will be called an *orifice*, the flow is dominated by dynamic forces. The passage cross-sectional area of the orifice is much smaller than that of the bearing clearance even, so V is much higher. The fluid has to accelerate to pass through, converting pressure energy into kinetic energy to do so (Bernouilli's equation encapsulates this process). Only a little of this kinetic energy is reconverted into pressure energy when the velocity falls after the orifice, that is, there is little *diffusion*: the rest is dissipated as thermal energy. The pressure drop, as Bernouilli's equation tells us, is proportional to V^2, and so to flow squared. Thus a 1 per cent increase in flow through an orifice results in a 2 per cent increase in pressure drop.

We could use as a restrictor a capillary hole, bigger in diameter than the orifice and much longer, in which the Reynold's number would be less than 2000 and the flow would be proportional to the pressure drop. Then an increase in flow of 1 per cent would cause an increase of only 1 per cent in pressure drop, and the bearing would be less stiff. On the other hand, a change in temperature will change the oil viscosity, and this will not affect the balance between a capillary restrictor and its bearing, whereas it will upset the balance between an orifice and a bearing.

There are two other ways of giving a hydrostatic bearing stiffness. One is simply to pump a given flow of oil through it. Then if h is decreased, since the flow remains the same, the pressure must rise. The other is to feed back the pressure in the recess so that when it rises it tends to open the restrictor and so reduce the pressure drop over it. It is an interesting design exercise to work out how to do this.

4.7 HYDRODYNAMIC BEARINGS: SQUEEZE ACTION

In a hydrostatic bearing, oil under pressure is forced between the bearing surfaces to separate them. If there is relative motion between the two surfaces, however, it may be possible to use it to generate the necessary oil pressure within the film. There are fundamentally two ways of doing this, by squeeze action and by wedge action. The less useful is squeeze action, but it is also the easier to understand, so it will be taken first.

Imagine that two equal circular plane surfaces are separated by a thick film of oil, when they are suddenly forced together, so that the distance h between them decreases (Figure 4.11). The oil is forced outwards, flowing between the two surfaces just as it did in the hydrostatic bearing, but clearly the flow is zero in the centre and a maximum at the outside. If W is the velocity of approach, the volume of the oil film within a radius r of the centre is decreasing at a rate equal to the area πr^2 times the velocity W. This flow $\pi r^2 W$ is passing through the circumferential gap at radius r, which is h high and $2\pi r$ wide, so the mean radial velocity V_m is

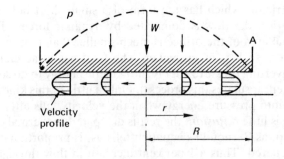

Figure 4.11 Squeeze action

$$V_m = \frac{\pi r^2 W}{2\pi r h} = \frac{Wr}{2h} \tag{4.6}$$

which increases with r. The pressure drop per unit length of radius is therefore proportional to r, so the pressure distribution is parabolic, as shown by the broken line in Figure 4.11.

It may be thought that this effect is not useful, because the velocity W can only be small, but then r is very large compared with h. With the additional result, proved in books on fluid dynamics, that the pressure drop per unit length in the *Hagen–Poiseuille* flow of Figure 4.10 is given by

$$\frac{dp}{dx} = \frac{12\mu V_m}{h^2} \tag{4.7}$$

the pressure distribution can be calculated and the mean value p_m of the pressure can be found. Now in equation (4.6), the velocity W is the rate of decrease of h, that is

$$W = -\frac{dh}{dt} \tag{4.8}$$

and it is readily found that under a steady mean pressure p_m the film will collapse from h_1 thick to h_2 thick in time t where

$$t = \frac{3\mu R^2}{4p_m}\left(\frac{1}{h_2^2} - \frac{1}{h_1^2}\right) \tag{4.9}$$

In practical applications, like little-end bearings, h_1 is several times larger than h_2 so the original thickness has little effect and we can put

$$t \simeq 0.75 \frac{\mu}{p_m}\left(\frac{R}{h_2}\right)^2 \tag{4.10}$$

An example

A rough calculation, of the kind designers should use frequently, can be made, based on equation (4.10), to assess the feasibility of using 'squeeze' action in a little-end bearing in a piston of a four-stroke IC engine.

This bearing, between the piston and the connecting rod, suffers several reversals of load direction in each cycle of two revolutions, and the clearance on the unloaded side can be kept full of oil, to provide a squeeze film when the load reverses. The severest load comes in the working stroke where the pressure in the cylinder is high. Will the film last long enough?

If the mean pressure in the cylinder on the working stroke is 1 N/mm², a typical figure, and if the projected area of the bearing (Figure 4.12), which is roughly a square 30 mm by 30 mm, is one-fifth the area of the piston, the mean pressure in the film, p_m, will be 5×1 N/mm² or 5 N/mm². Suppose the eventual thickness of the oil film, h_2, can go down to 8 μm without risk of failure, and the viscosity of the oil is 0.05 kg/m s, equation (4.10) yields

$$t = 0.75 \cdot \frac{0.05}{5 \times 10^6} \left(\frac{0.015}{0.008 \times 10^{-3}}\right)^2 = 26 \text{ μs}$$

where R has been put equal to 15 mm, half the side of the 30 mm × 30 mm square which is the projected area of the bearing.

This quick rough calculation has been included because it is typical in many ways of those a designer should make in the exploration of means. It is subject to all kinds of error, such as equation (4.10) being for a flat circular surface and not for a curved rectangular one (Figure 4.12), the mean effective pressure being averaged over *distance* and not *time*, and

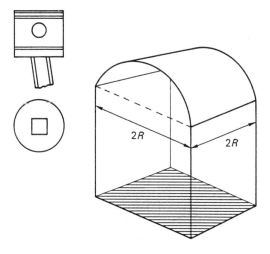

Figure 4.12 Projected area of little-end bearing

being variable in time at all, and the complete neglect of the inertia load from the piston (which averages to zero). The roughest figure of all is h_2, the final film thickness. This has to be sufficient to prevent the irregularities of the surfaces contacting one another locally over sufficient area to cause surface deterioration and the rapid ruin of the bearing, and will depend on the quality of finish and fit. Nevertheless, the figure of 26 μs for the survival of the film, corresponding to a speed of 1150 r.p.m. (when one stroke takes 26 μs) tells us that it is reasonable to expect a squeeze action bearing to work, provided we can guarantee oil in the clearance at the beginning of the working stroke.

In the case of the little-end bearing, it is well known that such a bearing will work and there is a wealth of experience (in two-stroke engines the load may not reverse, so special means must be adopted). But the designer will frequently meet problems outside the well worked areas, and then such simple rough calculations are an invaluable first step, enabling feasibility to be assessed and insight to be developed. Later, accurate values must be found by computing, but in the early stages rough answers will do and can be had without delay.

4.8 HYDRODYNAMIC BEARINGS: WEDGE ACTION

Most hydrodynamic bearings work by wedge action. Figure 4.13 shows the journal of such a bearing displaced in the clearance, the clearance being

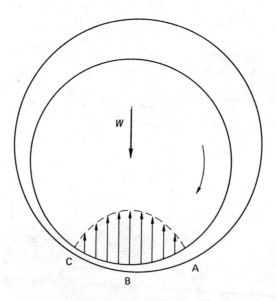

Figure 4.13 Hydrodynamic journal bearing

greatly exaggerated for clarity. As the shaft rotates, oil is drawn into the narrowing space at A and develops pressure which supports the journal: the upward-pointing arrows indicate the pressure in the film. Beyond B the pressure falls again. The term 'wedge' derives from the narrowing space ABC which generates the pressure.

The mechanism of pressure generation may be seen in general terms by reference to Figure 4.14, in which a wedge-shaped gap is shown between two surfaces, the upper moving left with velocity U and the lower stationary. Oil entrained at the right, at A, flows through and out at the left. Three velocity profiles are shown, at A, B and C: each has zero velocity at the lower, stationary, boundary and velocity U at the upper, moving, boundary. Since the area between each velocity profile and the vertical through A, B or C represents the flow at that point, these areas are equal. Because the width of the passage is greater at A than at B, and greater at B than at C, the profile at A must be hollow and that at C must be convex: B has been chosen to be that intermediate point at which the velocity profile is a straight line. The hollow curve at A shows that there is a pressure gradient rising to the left (as at A in Figure 4.11) and the convex curve at C shows a pressure gradient rising to the right. The pressure is a maximum at B, where there is no curvature of the velocity profile and so the pressure gradient is zero.

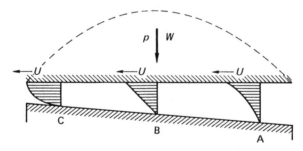

Figure 4.14 Wedge action: velocity profiles and pressure distribution

Thrust bearings can also be hydrodynamic, working by wedge action. However, unlike the journal bearing, support should be distributed all the way round the bearing, so a succession of wedges are used. One way of achieving this is the tilting pad bearing, another example of kinematic design (see Figure 4.15a). Here the stationary surface is composed of a series of tilting pads, each free to pivot about the point X, which is chosen so that the pad automatically sets itself to the right angle. One virtue of this is that when the oil warms the film thins, so that the angle should be reduced to keep the ratio of the clearances at inlet and outlet the same.

A simpler form of thrust bearing is shown in Figure 4.15b. Here a single step replaces the wedge. Below the step the velocity profile is like that at A

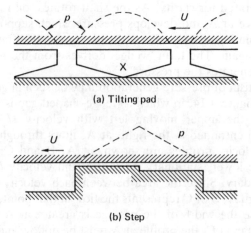

Figure 4.15 Thrust bearings

in Figure 4.14, above it the profile is like that at C, with only two values of pressure gradient, one up and one down, and a triangular pressure distribution like that shown by the broken line in the figure. A simple exercise is to think out how to design such a step thrust bearing to work in either direction of rotation. Step bearings are not often used, but the related pocket bearing is.

One difficulty with hydrodynamic bearings is that they rely on speed to develop the oil film. With heavy rotors such as those of steam turbine-generator sets in power stations, the bearings would be damaged on starting up before the speed was sufficient to develop an oil film: for this reason such bearings are often hybrids, working as hydrostatic bearings while starting up and switching to hydrodynamic operation at working speed.

The subject of film bearings has been discussed at some length, instead of simply referring the reader to engineering science text books, because it typifies the sort of thinking designers require. Before there was any scientific understanding of bearings, an ample journal would be provided together with a good supply of oil and some provision of holes and channels to get it where it seemed to be needed. Now the design of bearings is a highly developed technique, relying on well-understood processes and a wealth of knowledge from practice.

4.9 ROLLING-ELEMENT BEARINGS

Ball and roller bearings also rely on hydrodynamic films, very thin ones working at very high normal stresses (> 1000 N/mm^2), where the oil behaves more like a plastic solid than a viscous liquid. The load is carried by 'line' or 'point' contacts of the kind discussed in Section 2.7, and the capacity is limited by the contact stresses. These stresses are intermittent, with high frequencies of application, so fatigue under them is one major limit on load capacity. It is interesting to make a rough calculation of the load capacity of a roller bearing, taking the design maximum compressive stress to be f.

Figure 4.16 shows a cylindrical roller bearing. Let us compare its radial load capacity P with that of an imaginary plain bearing JJ of length B equal to the overall breadth of the roller bearing and diameter D equal to the pitch circle diameter of the rollers, by finding the nominal bearing pressure P would represent in such a plain bearing, that is, P/BD, denoted by p_E.

Following the treatment in Section 2.7, the contact patch between roller and inner track will be the middle cut out of an infinitely long ellipse, since R_A is infinite, that is, a rectangle of width $2b$ where

$$b = 4fR_B/E' \qquad (4.11)$$

and R_B is the relative radius of curvature at the contact.

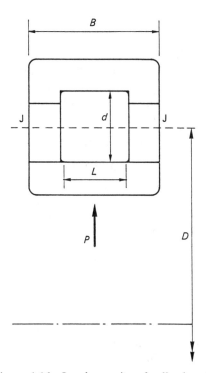

Figure 4.16 Load capacity of roller bearings

Now R_B will be somewhat less than the roller radius $d/2$, where d is the roller diameter. If we put

$$R_B = k_C d/2 \qquad (4.12)$$

then k will commonly be about 0.8, reflecting the effect of the track curvature. We have

$$b = 2k_C df/E'$$

If L is the effective length of the roller, that is, excluding the radii at the ends, then the area of a contact patch is $2bL$, and the average stress σ_{mean} will be $\pi f/4$. The load carried by one roller is thus

$$2bL\sigma_{mean} = \pi k_C dLf^2/E' \qquad (4.13)$$

The number of rollers N depends on their pitch p which is given by

$$p = \pi D/N$$

To allow for the cage, p has to be greater than d, generally by about one-third: we can represent this by another constant k_P, where

$$d = k_P p = k_P \pi D/N$$

so that

$$N = k_P \pi D/d \qquad (4.14)$$

Only a roller directly below the load will carry the full load given by equation (4.13). The rollers in the half of the bearing opposite the load will carry none at all, and those just into the loaded half will carry very little. Analysis shows that the average load is roughly one-fifth the maximum, so that the total load P is given by

$$0.2N \times \pi k_C dLf^2/E'$$

or, substituting for N from equation (4.14):

$$P = 0.2 \times \pi^2 \times k_P k_C DLf/E'$$

Now when allowance has been made for the cage and any shoulder, L will be a large fraction of B, that is

$$L = k_L B$$

so that we can write

$$P = 0.2\pi^2 k_P k_C k_L DBf^2/E'$$

To find the bearing pressure in the equivalent plain bearing JJ (Figure 4.16), we divide P by the projected area BD to give p_E, the equivalent nominal bearing pressure. Hence

$$p_E = 0.2\pi k_P k_C k_L f^2/E' \qquad (4.15)$$

Typical values of k_P, k_C and k_L are 0.75, 0.8 and 0.6, giving

$$p_E \simeq 0.7 f^2/E' \qquad (4.16)$$

Taking a typical value for f for the fatigue limit as 1450 N/mm and E' as 220,000 N/mm² gives

$$0.7 f^2/E' = 6.7 \text{ N/mm}^2$$

which will be found to fit well for most roller bearings up to inside diameters of about 200 mm. For larger bearings, p_E falls off rapidly.

It is instructive to consider why this simple result should come about. The contact patch area will be a fixed small fraction of the surface of the equivalent journal. If the patches are narrower, there will be more of them. This fraction is proportional to the ratio f/E' (see equation (4.11)), and the average stress it can carry is proportional to f, so that p_E is proportional to f^2/E'. Notice this has the same form as the expression for capacity to store energy that we saw in Section 2.10.

With a ball bearing, the value of p_E is much less. The length ($2a$) of the contact ellipse can only be a small fraction of the ball diameter, to avoid high slip velocities which result from the imperfect kinematics of a ball rolling in a groove where the arc of contact is large. The value of β (see Section 2.7) is 4, the mean stress is a smaller fraction of the maximum stress, and the contact patch is an ellipse instead of a rectangle: these last three effects combined reduce the value of p_E substantially. The overall effect is to give a value of p_E of only about 0.6 N/mm², less than a tenth that for a roller bearing.

This simple way of looking at rolling-element bearings gives a deep insight into the way the capacity varies with the size, and also leads us, for example, to expect the capacity of double-row self-aligning ball-bearings to be rather low.

5

Various Principles

5.1 INSIGHT AND ABSTRACTION

This chapter is concerned with some of the distinctive features of design thinking, a few of which have been touched upon already. This is a good point at which to emphasise the peculiar difficulties of design, or synthesis, as compared with analysis, which is more familiar to most of us and has almost a monopoly of that part of school education which is intellectually disciplined. One such difficulty is the need for a high level of *insight*, sometimes achieved only after long study, perhaps interrupted to allow 'incubation' to occur. This insight is the basis of that familiarity with the subject which may make original steps possible and is certainly essential to good design. Torroja, the great Spanish civil engineer, wrote that the designer should understand the working of a structure as clearly as he understood the working of a bow and arrow.

Another key to design is *abstraction*, which by stripping away the inessential, leaving only the essence, makes ideas easier to arrive at, even if they mostly prove later to be useless. A characteristic of innovative design may be the alternation between the abstract, generalised view and the concrete particular with all its reality and complexity. Ideas conceived in the abstract are fleshed out and found wanting, or refined and modified, in the concrete.

As a simple use of abstraction, consider the view of the radius of gyration of a cross-section as a measure of the spread-outness transversely of material and hence of its ability to resist bending. Simple as this idea is, it is a great help in form design. The steps involved are often the recognition that some part acts structurally as a beam, and that to increase its effectiveness in that role, its section must have a large radius of gyration. The designer then looks for sections that will have a large radius of gyration while also meeting other functional requirements.

5.2 BIASING

Biasing is perhaps the most general term that can be applied to an important abstract concept in design. Biasing is the superimposition upon some variable quantity of a large constant value in order to keep it within acceptable limits. Thus the input voltage to a transistor amplifier stage might range between ±3 V but the transistor will only work with a positive voltage. By applying a bias of +4 V the transistor can be kept in its working range. Similarly, a prestressed concrete beam, without the bias supplied by the prestress in the steel reinforcement, could crack on the tension side: the prestress swings the stress into the compression range, which the concrete can sustain. In some mechanisms, the loads in the parts might sometimes change sign, causing a small motion (lost motion or backlash) as the clearance changes sides (see Figure 5.1). By superimposing a biasing load, this motion can be avoided (lashlocking). Turbine blades suffer a bending moment due to the gas load on them: by inclining them slightly to the radial direction, the centrifugal load on them can be used to apply a bias to this bending moment, reducing its maximum value. In this case the object is not to maintain the sign of the variable always the same, but to reduce the maximum value. The technique is closely analogous to leaning into a strong wind when walking.

A bicycle wheel uses bias, in the form of prestressing, in at least two ways. The spokes are too slender to sustain load in compression, but by pretensioning them they can be made, in effect, to do so: the weight of the rider is transferred to the road via an increase in the tension in the spokes near the top of the wheels and a decrease in the tension in those near the bottom.

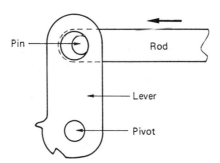

Figure 5.1 Backlash: because of the large clearance of the pin in the hole, the rod can move a little to the left before the lever moves (exaggerated)

The tyre also is prestressed by the air in it, and it is an interesting exercise to trace the path of the loads. Pumping up the tyre puts a tension in the beads, which resists the pressure-induced radial tension in the walls. However, where the tyre flattens in contact with the road the tension in the walls is decreased, and the unbalanced upward radial force exerted by the beads over the length of the contact patch supports the wheel. Thus the machine and its rider are supported by four little suspension bridges, in effect (Figure 5.2). The radial tension per unit length in the wall of the tyre is given, very closely, by

pressure × radius of curvature in the radial (meridional) plane

Where the tyre bulges above the flat, the radius of curvature, and hence the radial tension, is much less.

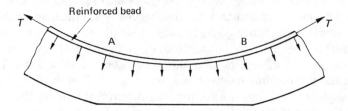

Figure 5.2 Bicycle tyre: the pressure produces radial forces in the walls, balanced by large tensions T in the beads ($\simeq 2$ kN). From A to B, the radial forces are much smaller, and the 'unbalanced' T components support the wheel like a hammock

Biasing, especially of forces, stresses or voltages, is a common practice, but it is important that the designer should recognise it even in very familiar circumstances. Thus it is often useful to think of bolts as preloading devices, as will be seen later.

5.3 FORCE PATHS

A very useful abstraction is the concept of a force path. Consider a bolt and nut and the 'path' of the force from the bolt to the support on which the nut rests (Figure 5.3). We can think of the force 'travelling' axially along the bolt and spreading out to the threads and hence into the nut, whence it 'flows' back to the face of the nut in contact with the support. If we zoom in on the threads, we can imagine the force gradually being shed from the male threads of the bolt on to the female threads of the nut as it 'flows' radially out into the nut.

The verbs 'travel' and 'flow' have been put in quotations because there is, in reality, no such thing as a flow of force, but the idea is useful, so long

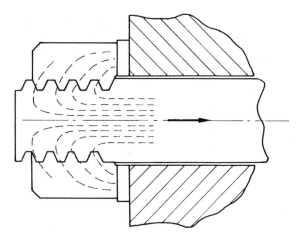

Figure 5.3 'Force paths' in nut and bolt

as we keep it vague: attempts to express it in precise terms are likely to end in fruitless complexities. Vague ideas are valuable when we are looking for suitable forms, provided always we can analyse the stresses later.

Principle of short direct force paths

It may seem obvious that force paths should be as short and direct as possible, but it is typical of the difficulty of design that this principle is often overlooked, and with it opportunities for better forms. The 'direct' path is often the vital one. Indirect paths often involve bending, which is an inefficient way of carrying load. Figure 5.4 shows the extra section needed

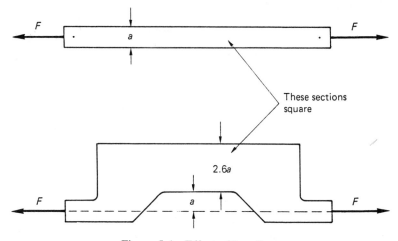

Figure 5.4 Effect of bending

to carry a load past an obstruction as against that needed with no obstruction. The original tie is square in section, $a \times a$, and to clear an obstruction which extends a distance a into the direct path without increasing the maximum stress, a section $2.6a$ square is needed, using more than six times as much material.

Figure 5.5 shows two ways of making a roller chain, such as bicycles have. Figure 5.5a is the usual way, with alternate pairs of links on the inside and outside, while Figure 5.5b uses dog-legged links, each on the outside at one end and the inside at the other. In both cases, the pins are subject to the two pairs of equal and opposite forces F, as in Figure 5.5c, producing a bending moment Ft in the middle section, where t is the thickness of the links. The straight links, however, are in pure tension, whereas the dog-legged ones are subject to bending moments $Ft/2$ on either side of the central joggle (see Figure 5.5d), which weakens the chain. It is interesting that in one of the first instances of a chain used as a machine element, in an ancient Greek automatic arrow-firing catapult, the links were joggled, to judge by the description: the chain was of wood.

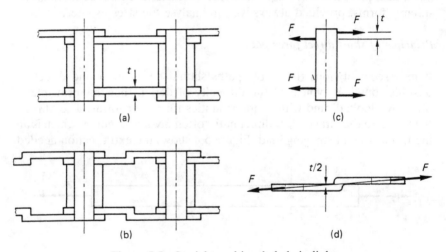

Figure 5.5 Straight and joggled chain links

5.4 A CYLINDER HEAD JOINT

This example serves to illustrate several of the previous points, and provides a good example of clarity of function. Consider a large CI engine and the means adopted to hold the cylinder head in place. Two functions are involved, that of preventing the cylinder head blowing off under the forces on it, and that of preventing leaks at the joint between the cylinder head and the cylinder block. It is convenient to think of these tasks as preventing *bursting* and preventing *leaking*, and the same pair will be met again. Sometimes it pays to separate these two functions, but not in this case.

The seal is usually provided by a gasket which is fairly flexible and is crushed by tightening the bolts so as to give a close fit with both surfaces. Each time the cylinder fires the resulting high pressure tends to force open the joint: whether it does so or not depends not only on the clamping force provided by the bolts or studs securing the head, but also on the flexibility of the parts between. If the pressure is able to force the surfaces of the cylinder head and the cylinder block apart between bolts, then the joint will fail however strong the bolts are. The cylinder block is a deep structure, very resistant to bending between bolts: the cylinder head has a bottom wall which may be too flexible unless it is backed locally by sufficient ribbing or other reinforcement, and this is just one of the many points that the designer must watch.

The other possibility which must be watched is that the whole cylinder head may lift sufficiently by stretching the bolts to allow gas to escape. This is prevented by prestressing, that is, by tightening the bolts till there is a big load in them, which also squashes the casings.

A similar situation, in the big end, was studied in Section 3.10, where it was shown that the bolts should be made as stretchy as possible. If k_B is the stiffness of the bolts and k_R the stiffness of the structural parts they clamp together, it was shown there that if a load G is applied to the whole, then this will be carried as G_B in the bolts and G_R in the structure, such that (Figure 5.6)

$$\frac{G_B}{G_R} = \frac{k_B}{k_R} \tag{5.1}$$

and hence

$$G_B = \frac{k_B}{k_R + k_B} G \tag{5.2}$$

In practice, k_B will be much smaller than k_R, so that G_B will be a small fraction of G, which is desirable to prevent fatigue, as in the big end. Nevertheless, it is desirable to make k_B even smaller by the means discussed in Section 3.10. This has the further advantage in the cylinder head bolts of reducing the effects of differential thermal expansion.

Once again, note that the abstract view of the bolt is as a clamping device which should have a low stiffness to perform its function well. If we were to take the study further and include the effect of the gasket, then the virtues of an absolutely low k_B would appear even more clearly.

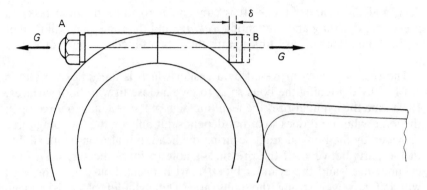

Figure 5.6 Load on big end joint

A common style of arrangement is shown in Figure 5.7. The studs clamp the cylinder liner between the cylinder head and the jacket or cylinder block. The clamping between the cylinder block and the liner occurs on quite a narrow land, ensuring a high pressure. This is a good practice: in the last century, working with lower quality machining, a lot of hand fitting and poorer materials, generous mating surfaces were usual.

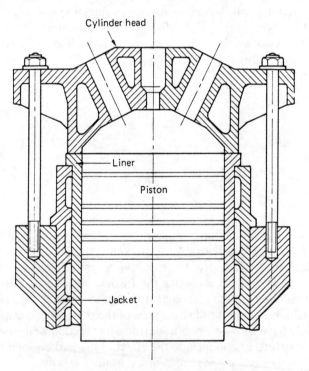

Figure 5.7 Diesel engine cylinder head

5.5 BEVEL GEAR MOUNTING

Figure 5.8 shows an example of clarity of function from a rig for testing high-performance bevel gears. The bevel gear G has just a short serrated shaft S1 on the back, and a flat face F. The rig contains a hollow shaft S2 running in bearings, particularly, a very large bearing B located just behind the gear. S2 has a flat face on the end which abuts the flat face F on the gear G and serrations in the bore which mate with the serrations on S1. A single large bolt through the axes of S1 and S2 tightens the face F against the face on S2.

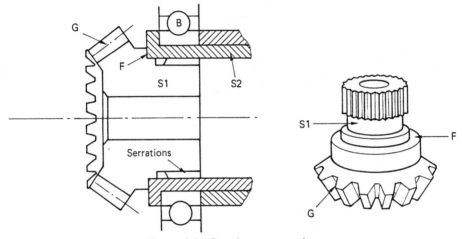

Figure 5.8 Bevel gear mounting

To change the gear, only the one bolt need be removed. The face F and the serrations are used as the datum in machining the gear, so accuracy of meshing is achieved without fitting. The serrations impose five constraints on G and the face F imposes three, so the location is slightly redundant (eight constraints instead of the ideal six), but this does not matter if the squareness of F and the serrations can be kept high. The very high loads forbid the removal of the two redundancies by making the serrations very short, since all the length is needed to carry the loads (see the discussion of the bush in Section 3.2). The heavy in-plane component of the load in the gear teeth is taken very directly by an amply-proportioned bearing close up behind the gear, with only a small reaction on the other, smaller bearing. All this is good, elegant design.

5.6 GEAR PUMP

In the discussion of the cylinder head, the difference between the functions of stopping leaking and stopping bursting was mentioned. Frequently it pays to make separate provision for the two functions, as in some gear pumps.

Figure 5.9a shows the principle of a gear pump. Two gears, with 8 to 9 teeth each perhaps, mesh together in a casing that fits them closely all round. One of them is driven and drives the other, and oil trapped in the spaces A is carried through from the lower pressure region (at the bottom in the figure) to the high pressure region at the top. A small amount of oil is actually carried in the reverse direction, in the gaps between the meshing teeth at B.

Everything depends on keeping the spaces A sealed and preventing leaks to the low pressure side. The parts shown in Figure 5.9a are sandwiched between two thick strong plates, and there is very little clearance between them and the end surfaces of the gear, as long as the pressure is not too high. Very serviceable pumps can be made on this principle. However, at high pressures the plates which are the bread of the sandwich bulge outwards, leaving a space between themselves and the end faces of the gears, and oil leaks back from the high pressure to the low pressure side.

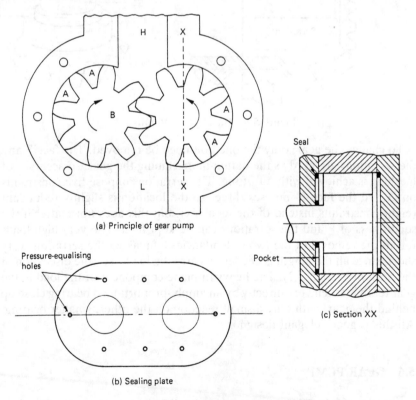

Figure 5.9 Gear pump

To avoid this leakage and so to enable the pump to work at high pressures, one way is to fit sealing plates on the sides of the gears (see Figure 5.9b), to make two more layers in the sandwich (the butter, if you like). It is aranged that these sealing plates are gently pressed against the gears by connecting pockets behind them to the oil flowing through the pump by pressure equalising holes (see figure). Each pocket is sealed off from adjacent ones, so that the effect is that of a number of patches of constant pressure loading on the back of the sealing plate, and the position of the holes and the extent of the pockets is chosen so that the net effect is a slight load inwards on to the ends of the gears.

The seals behind the sealing plates that separate the pockets must be able to accept a small movement without leaking, as O-rings can. The pressures in the pockets tend to bend the outer covers outwards, and it is this movement that the seals must be able to stand without leaking, which is just a matter of having enough 'squash' in them originally – see Section 10.4. Thus if the movement outwards is 100 μm, then an O-ring which is squashed by 200 μm will be adequate.

This technique of arranging for sealing parts to be nearly in balance, but just slightly forced against one another, is important and common. One example, the valve plate of a hydraulic pump, is treated at length in Section 9.2. The shaft seal shown in Figure 5.10 also works in this way. A ring of suitable material R is forced against a shoulder F on a shaft. There is pressure around the shaft to the right of the seal, and this pressure acts on a bellows of mean diameter D connecting R to the fixed casing. The pressure falls across the interface between R and F and forces R to the left. The pressure on the left-hand side of R between the outer diameter of R and D forces it to the right, and, by careful choice of the diameter D, the result is a small force pressing R and F together.

Characteristic of these examples is the way the designer has a clear view of how the parts should function, and arranges the shapes and chooses the dimensions so that they shall behave in just the way he determines.

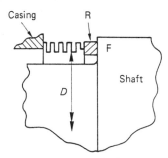

Figure 5.10 Shaft seal

A cardinal rule of design is to avoid arbitrary decisions (Chapter 1). In practice, the designer usually has to make *temporary* arbitrary decisions in order to move ahead, but he should always review them later.

5.7 NESTING AND STACKING

An important decision often faced by the designer is that of nesting or stacking order. A familiar example is provided by typewriters and some printers. Various motions have to be given to the type and to the paper in order to bring the right character to the right place on the paper and finally to 'strike' the two together to print the character. Various *nesting orders* can be used to achieve this. For example, for several decades most typewriters had a separate piece of type (type bar) for each letter (a and A, say) operated by a separate mechanism: such typewriters gave two motions to the paper, whereas now the paper usually has only one motion.

The motions involved are as follows:

move along the line	(feed)
move down the page	(scroll)
select characters	(selection)
select upper or lower case	(shift)
print the character	(strike)

The old typebar typewriter had a stationary frame relative to which the paper had two motions, feed and scroll, scroll being nested outside feed, that is, the whole scroll mechanism moved as the feed was operated. On the other side, the entire key mechanism moved when shift was operated, and the key mechanism provided both selection and strike in one movement. This arrangement can be represented as in Figure 5.11.

A slightly more abstract view is as follows. The motions have to be placed in order between the two components, type and paper, whose relative movement is to be controlled. At some point in this order there must be a fixed frame or 'earth', so that the old typewriter can be represented as:

type – (selection, strike) – shift – *earth* – feed – scroll – *paper*

Where there is no combination of function, such as 'selection' and 'strike' combined in one motion, then there are six functions to be ranged between the ends, type and paper, giving 6! or 720 possible orders. In practice, nearly all these orders can readily be seen to be useless, and only a few are worth considering. 'Golf ball' and daisywheel typewriters are

VARIOUS PRINCIPLES

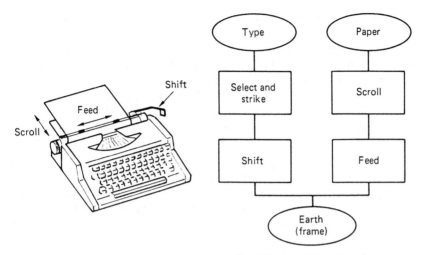

For shift, the whole basket of keys, linkages, typebars etc. moves

Figure 5.11 Nesting order of old typewriter

type – strike – select – shift – feed – *earth* – scroll – *paper*

with only the 'scroll' motion given to the paper.

Machine tools have nested motions. In a lathe, if we consider just turning, axial feed and cross feed, the representation is

tool – cross feed – axial feed – *earth* – turning – *work*

Consider a few other machine tools. Are there alternative nesting orders worth considering?

An interesting case is that of the front wheels of vehicles: relative to the body (which is 'earth' in this case), these wheels have three motions, rotation, steering and suspension, in this order:

wheel – rotation – steering – suspension – *body*

However, motorcycles have

wheel – rotation – suspension – steering – *body*

There is a general principle of great importance in suspension design, which may be expressed as:

'reduce the unsprung weight'

(or, more properly, mass). The strict application of this principle leads to the nesting order of the motorcycle, not that of the car.

5.8 GUIDING PRINCIPLES FOR CHOOSING NESTING ORDERS

It is easy to find guiding principles which assist in the choice of good nesting orders. For instance, it is usually desirable to put the earth somewhere near the middle, so that the nesting order can be regarded as having two branches starting from earth, as it is the number of motions in one branch which leads to complexity and clumsiness. Thus in Figure 5.11, the old typewriter had two motions in the type branch and two in the paper branch. The new typewriter has four motions in the type branch (strike, select, shift, feed) which is rather a lot, but is made more acceptable because two of the motions are very small (strike and shift).

One great advantage of the new typewriter is its small 'footprint' (area required to stand on) which is achieved by giving the feed motion to the type rather than the paper. The minimum possible width occupied by the old form was twice the width of the paper, while that of the new is the sum of the width of the paper and the width of the type head (see Figure 5.12) which is only about two-thirds as much. This gives us another principle, 'give large motions to small parts'. This practice also tends to reduce the size of motors or actuators and inertia forces, and even the cost of guides and structure.

To see another principle, consider the simple nesting order of the lathe:

tool – cross feed – axial feed – *earth* – turning – *work*

and consider the possibility of reversing the order of the two feeds. Figure 5.13 is a sketch of such an arrangement and its clumsiness is apparent. The longitudinal guideways are mounted on a moving table that has a short travel across the bed. There is nearly as much machinery and material in the longitudinal guides and much more in the cross-slide.

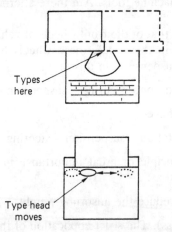

Figure 5.12 Comparative widths of typewriters

VARIOUS PRINCIPLES   113

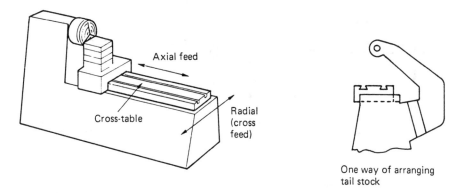

One way of arranging tail stock

Figure 5.13 Lathe with radial and axial feeds reversed

The stiffness is bound to be less unless a lot more material is used, because of twist in the 'cross-table'. This could be overcome by a multiplicity of dovetails, but it is difficult to make and fit such a guide accurately and it conflicts with the principle of least constraint. Also, the tailstock and any steadies required cannot be mounted directly on the bed, but must be overhung from the back, an arrangement which is expensive and lacking in stiffness. Also, very short, wide guides suffer from a problem all designers should beware of, the 'sticky drawer' problem (Figure 5.14a). Here the drawer, which is short and wide and has its two handles set well apart, jams when an attempt is made to pull it out using one hand.

The sticky drawer

What happens is that the drawer turns slightly (the figure is greatly exaggerated) so it touches only at diagonally opposite corners, A and B. The drawer is subject to three forces, the pull P and reactions R_A and R_B at A and B. If the drawer is sliding, then R_A and R_B will make an angle λ with the normals to the guides, where λ is the angle of friction.

Let the lines of action of R_A and R_B intersect at C. Now if the line of action of P passes to the left of C, say, cutting AC in D, the drawer will jam. For the forces P, R_A and R_B could be in equilibrium if their lines of action all passed through D, and the line of action of R_B would then be BD, which is at less than the limiting angle of friction λ, so that no sliding could take place. Detailed analysis would show this rigorously, but it is enough to recognise that in the statics of friction Murphy rules: if it can stick, it will.

The sticky drawer has more to teach the designer. For clarity, the drawer in Figure 5.14a is shown as a very loose fit. If it were a very good fit, things could well be much worse, for the reactions might not occur at the corners, but nearer the centres of the sides, and that would make C nearer the

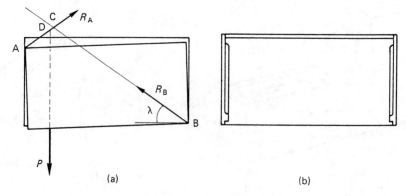

Figure 5.14   The sticky drawer

middle, and the distance P could be from the centre would be less. The lessons we draw are that guides should be long and surfaces corresponding to the sides of the drawer should be relieved except at the corners (Figure 5.14b). In machine tools, however, guides are not relieved in this way because it would increase wear and reduce stiffness. Another point to note is that guides should be long, not only to reduce friction, but to maintain squareness: in Figure 5.14a, if the clearances were the same but the drawer were twice as deep, then the angle of cant would only be half as much.

5.9   OTHER ASPECTS OF NESTING

Another aspect of nesting is the problems arising from driving a motion one or two removed from earth. For example, in the lathe, the cross-feed has to be driven from the bed. Because the axial feed is between the cross-feed and the bed, a drive is required which can operate independently of the relative axial movement of the cross-feed lead screw and the bed. This is achieved by a shaft running the length of the bed on which a part in the carriage is free to slide longitudinally but not to rotate. The rod and the part can have any non-circular section but in practice it is usual to provide a circular feed rod with a single groove in it, with the part in the carriage to fit, as in Figure 5.15. Where a rotational motion separates two components, difficulties may also be found. Sometimes the only way round this is use two concentric shafts, one about which the rotation takes place and a drive shaft running through the centre of it.

Wheel suspension

Where motions are sufficiently limited in extent, there are other possibilities. In a front wheel drive car, the up-and-down straight-line (roughly)

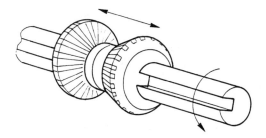

Figure 5.15 Cross-feed drive

motion of the suspension and the rotary motion of steering both separate the rotation of the wheel from the engine that produces it. In a wishbone suspension the wheel carrier H (Figure 5.16) forms part of a four bar chain A, B, C, D. Steering is by rotation about the axis BC, B and C being ball joints while A and D are long hinges attaching the wishbones to the body. The shaft driving the wheel, E, F, G, has joints at E and F so that it can flex to accommodate the up-and-down motion of the suspension and the rotation of steering.

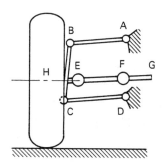

Figure 5.16 Front suspension

It is instructive to count degrees of freedom. Because the hinges at A and D are long, each of the points B and C has only one degree of freedom relative to the body, motion in a circle about A or D respectively. Thus the wheel carrier H has two points, B and C, whose degrees of freedom have been reduced from three to one, that is, four degrees of freedom have been removed, leaving two of the original six: these two are steering and the suspension movement. If E and F were ball joints, EF would be a further constraint, with redundancy and conflict between AB, EF and CD. To avoid this, a further degree of freedom must be introduced: this could be a splined joint (Figure 5.17) in the shaft, but more usually now the joint F is made to slide ('plunge') as well as bend. Readers may like to return and carry out the exercise of counting freedoms on the suspension of Figure 1.8c.

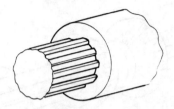

Figure 5.17   Splined sliding joint in shaft

Figure 5.18 shows a constant velocity joint of the most common (Rzeppa) type at (a), and a 'plunging' joint of the GKN Birfield sort at (b). The constant velocity (CV) joint is used at E, and is a ball joint in which opposed meridional grooves in the ball and its socket embrace balls which prevent relative rotation about the axis of drive. The plunging joint is more complicated and has three interdigitated prongs on each half, with balls between them.

Note that both of these joints have the vital property that the velocity ratio between the input and output is constant, unlike the universal or Hooke's joint which has fluctuations in velocity ratio at twice shaft speed and would be quite unacceptable at these angles of bend in the shaft and for this application.

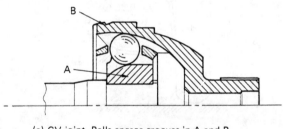

(a) CV joint. Balls engage grooves in A and B

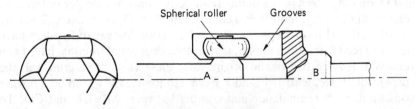

(b) Plunging joint. Three spherical rollers running on radial arms on A engage in three axial grooves in B

Figure 5.18   Constant velocity joints

The plunging joint with its extra degree of freedom has a more difficult specification, so it is put at the inner end, at F, where the flexure is less. The result is two difficult design problems, whereas if they were reversed one problem would be easier and the other very likely impossible: this kind of situation is common in design.

We need to consider also the interaction of the other two motions, steering and suspension. Because the steering is 'outside' the suspension it does not affect it, except by contributing undesirable extra unsprung weight. On the other hand, the steering linkage has to bypass the suspension motion, which it does by the same means as the drive to the wheel, by joints which add degrees of freedom. However, complete independence is practically impossible. Deflecting the suspension will in general produce a small change in steering angle, but the amount can be kept acceptably low by a careful choice of geometry.

In the problem of drives to motions not adjacent to earth we have a further aspect of nesting which may point in the opposite direction to the guide, 'put large motions nearer to earth'. The jointed shaft solution of the suspension problem would not work for the cross-feed of the lathe because of the length of the motion in between. Notice that rotary motions are particularly difficult because electrical connections, while accepting a limited rotation, cannot survive an indefinite number of revolutions unless those highly undesirable components, slip-rings, are used. Otherwise, electrical drives are often a good way of avoiding the transmission of motion through several tiers of the 'nest'.

Surprisingly, it is possible to connect a steadily rotating body to a stationary one by a cable, without slip-rings and without the cable becoming twisted up and breaking, but this fact is useful only in a very few special applications, such as the making of cables themselves. This seems totally in conflict with common sense, but it is true. To prove it, take an electrical kettle with lead attached but the plug not inserted in a socket. Point the plug end of the lead straight at the kettle end, so that the form of the lead is like a question mark, as in Figure 5.19. It will be found that the plug can be rotated indefinitely about the axis xx, a hundred times if you like, without the lead becoming twisted. For every second turn of the plug,

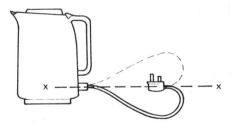

Figure 5.19   Kettle experiment

however, the lead must orbit once round the axis xx, as shown by the broken line. Because the plug is completely enclosed in the orbit, it cannot be held by one hand all the time, a feature which accounts for the limited use of this piece of kinematics.

5.10 SUMMARY OF GUIDING PRINCIPLES FOR NESTING ORDERS

These are some of the guiding principles which can help the designer in deciding on a nesting order:

(a) earth should be near the middle of the nesting order;
(b) branches should be as many and short as is practicable;
(c) lightest, smallest, fastest and most frequent motions should be furthest from earth; and
(d) heaviest, largest, slowest and least frequent motions should be nearest to earth.

To these might be added, prefer pivots to slides for outer motions and flexural elements to both for small motions (see Section 5.11).

It is interesting to apply these principles to some of the wider range of engineering devices that exhibit nesting orders.

A few examples are metrology instruments like the Talyrond, helicopter rotor heads, telescopes, milling machines, robots, welding manipulators and cranes. Question familiar designs, for instance, are we sure that cars should have the steering outside the suspension? and follow through the consequences. Should the headlights be attached to the steering in a 'suspension outside steering' car so as to 'look round the corner'? Examine the ways used to cope with internal relative motions, like electrical connectors that are made as wide flat strips to accommodate small radii of curvature when bent or rolled.

5.11 FLEXURAL ELEMENTS

The achievement of quality in engineering products is greatly advanced by eliminating potential defects, and these exist wherever one part rubs on another under any substantial load or at any substantial speed or frequency. Lubrication is needed and may not be provided or may fail or be contaminated, corrosion or vibration may cause fretting, and so on. Moreover, friction between the surfaces and necessary clearances causes undesirable behaviour, such as lost motion and hysteresis. For these reasons there are great advantages in 'solid state' design, achieving relative motion by bending or twisting solid elements.

A relatively old but very instructive example is provided by suspension bridge towers (Figure 5.20a). It is good design to balance the loads T_1 and T_2 from the cables. This has been done by providing a kind of pulley at B to carry the cables over the top (actually a saddle on rollers, Figure 5.20b). Sometimes the tower has been hinged at the bottom. Nowadays the tower is usually made sufficiently flexible to bend enough to make T_1 and T_2 nearly enough equal. This gives rise to a problem of disposition of stress (Chapter 1). The design stress in the material of the tower must be shared out between three major uses: the direct stress due to the weight of the bridge, the stress due to the transverse wind loading on the bridge, and the bending stress that comes from the compliance of the tower to the cables on either side as they adjust to one another. The stress due to wind loading will be largely bending, if the tower, viewed along the bridge, is a Vierendeel truss (Figure 5.20c), largely direct if it is a braced truss (Figure 5.20d). If σ_D is the direct stress, σ_W the stress caused by the wind and σ_F the stress caused by flexing to comply with the cables:

$$\sigma_D + \sigma_W + \sigma_F = f$$

(where the stresses will all be compressive in the worst corner). A simple optimisation will give roughly the best ratios of the σ terms to one another, that is, what fraction of the allowed stress should be reserved for each of the three functions. The complicated final calculations would modify these ratios slightly.

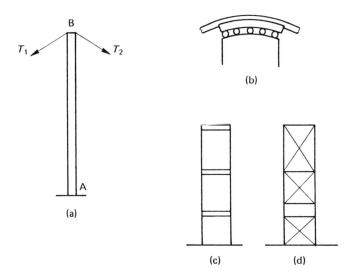

Figure 5.20 Suspension bridge towers

The Vierendeel truss resists shear by bending in its chords and columns (here, unusually, the chords are vertical and the columns horizontal). This is very inefficient structurally, and so may offend a knowledgeable eye. However, it is used with justification in the ladder frame chassis of trucks (the term explains itself) where the need for unobstructed spaces is great.

Vehicle suspensions often contain links which should ideally have ball joints at the ends, but which move through only a small angle. In such cases it may be possible to fit a rubber bush instead of a ball joint, removing the need to lubricate and reducing the transmission of high-frequency vibration and noise. However, rubber bushes are soft transversely as well as angularly, so that they reduce the stiffness of the constraints and hence the precision of the suspension geometry and the steering.

An impressive use of flexural elements was the basis of the first comparator which was able to measure a difference in diameter between two cylinders of only one-millionth of an inch (0.025 μm). In Figure 5.21, A and B are two smooth hardened steel blocks. A is fixed to a stand and B slides on A. Brazed to A and B are two very thin steel strips united by a rather thicker piece C at the top. When B moves upward a small distance δ, the piece C rotates and a long pointer attached to C moves over a scale. If t is the distance between the two strips, the magnification of the movement is L/t, where L is the length of the pointer, and this ratio can be made several hundred times by making t very small and L large.

Effectively, this mechanism is simply a lever, with one arm of length t and one of length L. It would be very difficult, however, to make a lever work smoothly with such a high ratio.

This is not quite a 'solid state' comparator because there is sliding between A and B. This can be eliminated by mounting B on A by means of

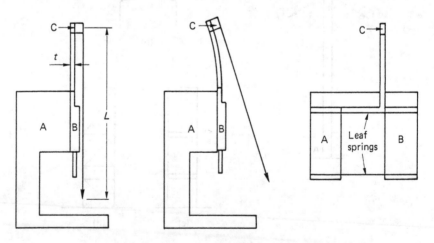

Figure 5.21 The millionth comparator

two further, more robust, parallel steel strips, after the same fashion as was used in the bathroom scales in Chapter 2. Note that unless these strips are nearly straight and the movement of B is small, the movement of B may bring it closer to A sufficiently to upset the geometry of the measuring system. By making A a tube and mounting B on diaphragms within that tube, a straighter motion of B is obtained.

Figure 5.22 shows the principle of a flexural pivot of a type used in delicate but heavy instruments. An arm carrying a large mass (not shown) is effectively pivoted at O, where two thin steel strips PQ, SR, are encastered in a support at P and S and in the arm at Q and R. The load *mg* due to the mass keeps the strips always in tension and so anchors O firmly, but rotation is possible and only weakly resisted by the flexing of the strips.

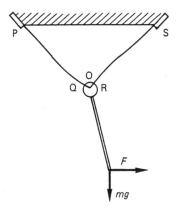

Figure 5.22   Flexural pivot

A conflict may arise if it is desired to make the centering of O very positive, that is, to make it very stiff laterally: to achieve this, the strips could be made thicker, but that would increase the rotational stiffness. An ideal pivot would have zero rotational stiffness and infinite transverse stiffnesses. It would also be free of any slack and quite frictionless, and possibly able to take high loads and run at high speeds.

5.12   EASEMENTS

Designers are often very quick to recognise the special difficulties of a job, but they do not always spot the ways in which it is easier, the 'easements', to use an old word where there does not seem to be a modern one. The specification of a bearing, for instance, usually has easements.

A machine tool bearing has to be free from clearance, and also very stiff, so that it does not give under load and impair the accuracy. Antifriction bearings have appreciable elasticity under light load, before large contact patches develop (Chapter 2), so they may need to be preloaded. On the other hand, a fair amount of friction can be accepted.

A bearing in an instrument may have to be as nearly frictionless as possible and as stiff as the bearing in the machine tool, but it may have to rotate only through a small angle, making a flexural pivot a possibility, and it probably only needs to move slowly. Though it has to be stiff, the loads on it may be low. The requirements for a few typical cases have been set out in Table 5.1: V indicates a demanding requirement, M a moderate one, and a blank an insignificant one or non-applicability. Note that no case has more than three Vs out of a possible five.

Table 5.1 Requirements for pivots

	Headstock	Vehicle hub	Wind tunnel balance	Gas turbine	Roundness measuring M/C	Weighing M/C
High load capacity	M	V	M	V	—	V
High speed	M	M	—	V	—	—
Small clearance	V	—	V	—	V	—
Low friction	—	—	V	—	V	V
Transverse stiffness	V	M	V	M	V	M
Unlimited rotation	Y	Y	—	Y	Y	—

M: Moderate demand; V: Extreme demand; Y: yes.

This chapter has explored a little of the complex field of design. The next ventures on a much larger area, even more complicated and impossible to map in the linear fashion a book demands, that of materials and manufacture.

6

Materials and Manufacturing Methods

6.1 INTRODUCTION

The joint influence of materials and methods of manufacture on form design is profound. Most of the elements in a design are likely to have a structural function, and so the mechanical properties of the material, its strength or its stiffness, will determine the *scantlings*, the general thickness of the form. Thus in cruise ships, there is a premium on passenger space, and this can be enlarged by adding extra decks, a process which is limited by the need to preserve stability to prevent capsizing. The use of aluminium alloy instead of steel for the upper decks reduces the weight of them, and so increases the space which can be provided. The aluminium alloy is weaker than steel, so that the scantlings must be increased by about 50 per cent: as the density is only about one-third that of steel, however, the weight is reduced to about one-half. Just on the weight-saving aspect, it looks a good idea to use aluminium alloy.

However, the question of materials is confounded with that of manufacture. Ships are built by fabrication from plates and rolled sections welded together. This can be done with aluminium alloy, although the welding is more difficult. Then there is the matter of cost, which is much higher than for steel, and so it is a matter of calculating whether over the life of the ship the extra space will yield enough extra income to justify the higher initial cost. There are many other aspects, too, such as corrosion, especially if steel and aluminium are to be used together in one structure exposed to the action of air and salt water.

A lawn-mower

At an early stage in design it is usually necessary to decide on the type of material, and often the type of manufacturing process too, that is to be used for the major components. Often the choice is very restricted, largely by economic factors. The working parts of a lawn-mower, for instance, are mostly made of steel because anything else which would meet the functional requirements is much more expensive. Casings or covers on the mower may present more choice, between steel pressings or glass-reinforced polymer perhaps, and from that choice will spring all kinds of consequences, some of which even the experienced designer may overlook.

If the mower has a petrol engine, the running parts will mostly be of steel or some superior, non-brittle, cast iron (nodular or spheroidal-graphite iron). Cast iron is preferred to steel for many parts because it has nearly equal properties and can readily be cast, which is cheaper than forging or machining from solid. The major stationary parts of the engine will probably be cast in aluminium alloy. These castings are quite complicated, and so the greater ease of casting, and the good finish if die-casting is used, as would normally be the case, favour aluminium alloy. On top of that, as in the cruise ship, there is a valuable saving of weight.

The relationship between design and production

Later in design, in the embodiment stage, there are more detailed choices to be made, such as which steel is to be used, and just how is it to be given the required form. However, at this stage, the reverse process should dominate: exactly what form should be chosen for this part so that, besides meeting the other functional requirements, it can also be made economically? On the whole, in the conceptual stage materials are chosen to suit the design, and in the embodiment stage the design is chosen to suit the material.

It is often said with justice, that the aspect of designing to suit the material and the manufacturing method is not sufficiently pursued in most design offices. To some extent, this is a management problem, and hinges largely on arranging the right form of discussion between the designers and the production people at an early enough stage. However, it is very important that the designer should be fully alive to production considerations and sufficiently knowledgeable to understand the implications for manufacture of his decisions. Above all, he should have a good idea of what can be done, how difficult it is, and about what it will cost, and make a pride of meeting the production man more than half-way. To do this requires him to keep abreast of new developments and the implications they have for design, which are often limited in field, but dramatic in their effect in the right application.

6.2 MATERIALS, MANUFACTURE AND DESIGN PHILOSOPHIES

Manufacturing processes may be categorised as *generative* and *replicative*. Replicative processes are those like casting and moulding and forming in presses where the form is in the tools used. Generative processes, often referred to in everyday work as 'knife and fork' methods, are those like machining where the form is generated by relative movements of tool and work piece, as a lathe generates cylindrical surfaces, cones and flat faces.

In general, replicative processes have higher first costs and lower marginal costs than generative ones. The approximate total cost of producing n items can be written in the form

$$A + Bn \qquad (6.1)$$

where A is the cost of tools, dies, fixtures and so on which are incurred before even one part can be made, and B is the further cost of making one off, the material, the costs of the machine and its operator, the rent on the space occupied and so on (it is assumed that the machine would be engaged on other work if not making the parts in question). Then the cost per item is

$$C = \frac{A}{n} + B \qquad (6.2)$$

with the characteristic flattening hyperbolic form of Figure 6.1. In the generative process A is usually relatively low and B relatively high compared with a typical replicative process. It is clear that large values of n favour replicative methods of manufacture.

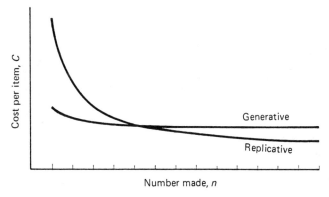

Figure 6.1 Typical variation of cost with number made

A few recent tendencies in engineering have favoured smaller volumes of production and a greater diversity of products. Powerful movements in other fields, however, such as the development of product liability law to provide rich pickings for enlarged flocks of lawyers and the ever-rising and more invasive tide of bureaucracy, will probably result in fewer products being made in greater numbers. For this reason, replicative methods are likely to dominate, so dictating a design philosophy which is attractive in itself and desirable for the preservation of the environment.

In many cases, the accuracy of replicative processes is inadequate, so that the appropriate technique is to use a part largely made by replicative methods but finished by generative ones. To achieve designs which can be made in this way is a stimulating challenge to the designer. An advantage of replicative processes is that they usually produce little scrap material, that is, they are 'near net form' processes. This keeps down the material costs, avoids waste of scarce resources, saves energy used in reprocessing material, and is generally a good thing.

The appropriate philosophy is therefore often to design for replicative production methods, supplemented by generative sizing and finishing operations. A simple example is shown in Figure 6.2. The part shown at (a) is made by upsetting a piece of 6 mm mild steel rod, as at (b), turning down the end as at (c), and flattening the end as at (d). Finally the slot is punched in the flat end.

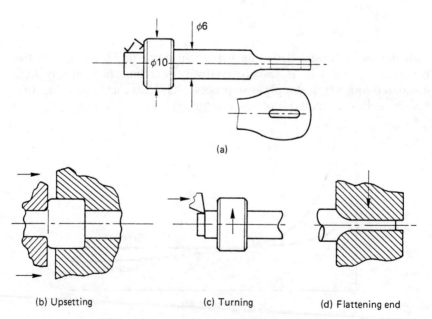

Figure 6.2 Formed part with local machining

For a one-off item, for a test rig, say, the part could be turned from a solid rod of diameter 10 mm, making a greater weight of swarf than part.

Some other limitations of replicative methods need to be recognised. For instance, castings cannot always be made as thin as the function would allow: the designer may have to accept a wall thickness of 4 mm when 2 mm would be quite strong enough. Sometimes an alternative is to use a fabrication from several pressings, when a thin uniform wall may be obtained. However, a great advantage of castings is that local features such as bosses or lugs can be added at negligible extra cost.

The inferiority of fabrication in this respect is well shown by the case of bolted joints, where a casting can be thickened locally but a fabrication may require to be spot-faced, reducing the thickness just where greater thickness is desirable to diffuse the large local load into the structure (Figure 6.3).

The most striking uses of replicative processes are perhaps in the polymers, where the low moulding temperatures make the dies very stable in use and thin sections and sharp features are readily made, advantages which can be exploited in snap fitting assemblies and elaborate ribbing and stiffening. Zinc-based casting alloys share these advantages to some extent, notably in competition with aluminium alloys.

Steel and cast iron, aluminium alloy, zinc-based alloys and polymers provide us with a spectrum of castable or mouldable materials, decreasing in strength but increasing in facility and precision of casting, and all of moderate cost. By adding glass fibres to the polymers, their stiffness and strength may be increased, but with a reduction in ease of moulding. For most parts of high or middling solidity and other than simple form, these materials will usually supply the solution finally chosen. Cases of low solidity and some machine parts will be discussed later.

The division into replicative and generative methods of manufacture is not sharp. For instance, pressure vessels are often made by the automated winding of filaments on a mandrel, a process which has elements of both.

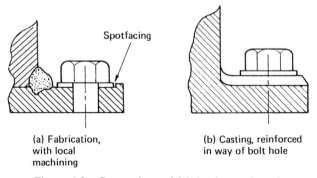

(a) Fabrication, with local machining

(b) Casting, reinforced in way of bolt hole

Figure 6.3 Comparison of fabrication and casting

6.3 THE EFFECT OF DENSITY

Many materials have about the same specific strength, where

$$\text{specific strength} = \frac{\text{ultimate tensile strength}}{\text{density}}$$

(It would be better to use a design stress of some kind, but ultimate tensile strength is usual.) For example, to carry the same tensile load would require a tie of about the same mass, whether it were made in wood, aluminium alloy or steel. Similarly, the specific stiffness, where

$$\text{specific stiffness} = \frac{\text{Young's modulus}}{\text{density}}$$

is about the same for many materials, including these three. Thus if we express their structural properties per unit mass, these very different materials appear very comparable. However, when specific strength and specific stiffness are roughly equal, low density is a great advantage in some cases, and a disadvantage in others.

Low density is a great advantage where elastic stability is important, which is generally in structures of low solidity, while high density is an advantage when space is very restricted.

An example of a low solidity structure is a slender column. Here, low density gives a greater cross-section of material and hence a greater flexural stiffness for the same mass per unit length. Figure 6.4 compares the cross-sections of slender solid cylindrical columns of steel, aluminium alloy and wood to carry the same load, assuming relative densities of 3, 1 and 0.2 and Young's moduli in the same ratios. Although more slender, the steel column weighs 1.73 times as much as the aluminium one, which itself is over twice as heavy as the wooden one. Of course, by using hollow sections the masses could all be reduced, but still the lower density

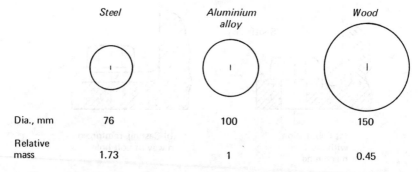

Figure 6.4 Slender columns to carry the same load over the same length

materials would continue to show an advantage until the walls became too thin to be practical.

In contrast to the column, Figure 6.5 shows the cross-section of cylindrical beams in steel and aluminium alloy to carry the same bending moment and to have an outside diameter of 100 mm. The relative densities are 3 and 1 as before, and the allowable stresses are in the same ratio. This time, because of the restricted outside diameter, the steel comes out much lighter.

There is a final advantage of low-density materials, which is that even with replicative production processes, there are regions of useless material which it is not practical to save by complicating the form: the lower the density, the less relatively is the waste, and the less the unnecessary additional mass. Density is just one of many properties which need to be considered in choosing materials.

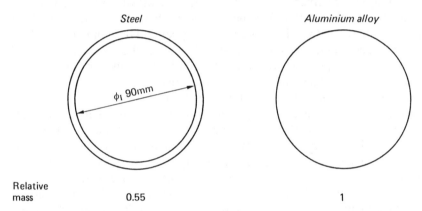

Figure 6.5 Sections of diameter 100 mm to carry the same bending moment

6.4 CHOICE OF MATERIAL

It is convenient to think of the properties of materials which affect the choice of the designer under three heads, functional, environmental and manufacturing. The first of these, functional, can also helpfully be divided into two, structural and other. A brief list is as follows:

Functional Properties

Structural	*Other*
Strength (of all kinds)	Conductivity (of heat or electricity)
Stiffness	Ferromagnetism
Density	Lubricity
Dimensional stability . . .	Transparency . . .

Environmental Properties
Heat resistance
Resistance to corrosion
Resistance to ultra-violet . . .

Manufacturing
Mouldability and formability
Machinability
Exotic properties, like superplasticity
Amenability to surface treatments . . .

This is a very short and simple list, suitable for use as a check list. For the springs of the bathroom scales for instance, sheet spring steel meets all the requirements, except for a possible question mark about resistance to corrosion, and whether a cheap surface treatment will suffice or not.

Are there other materials which would be better than steel for these springs? There are two kinds, beryllium–copper and titanium alloys, which would be about as good as spring materials, but much more expensive, and it might also be possible to use a fibre reinforced polymer, which again would cost more. The difference in cost means that other aspects, such as variations with ambient temperature, need not be gone into.

However, we should ask the question, could one of these other materials confer any other benefits which would justify their extra cost? There is a sense in which the answer can never be a confident 'no', because we cannot be sure we have considered every possibility. Here is an example of a way in which it might be feasible to use a composite to advantage.

A weakness in the bathroom scales is the clamping of the springs at the ends (see Figure 2.12b). To ensure the spring is tightly encastered, it must be firmly clamped. In these circumstances, clamps and springs tend to behave as one continuous piece of steel, with a high stress concentration factor which reduces the allowable stress. Moreover, however much the joint is tightened, there must be a tendency to slip between clamping faces and spring, which will introduce a slight hysteresis and result in inaccuracy. One way of avoiding this would be to thicken the springs at the end, as in Figure 6.6, but this would have to be done either by machining away most of a thicker sheet or by forging, both expensive. Figure 6.7 shows a composite spring consisting of two similar T-shaped blocks with composite leaf springs filament-wound on them at F. The blocks B1 and B2 would be mouldings of the same basic material, with generous flat faces for attachment directly to the base and platform respectively. A problem might arise with the bonding at the points P. The efficiency of use (Section 2.10) would be low, at one-ninth compared with one-third in the steel springs with their diamond-shaped cut-outs, but the superior energy storage of the polymer would compensate for this. But overall, the neat

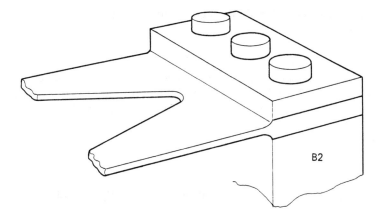

Figure 6.6 Spring with thick ends

one-piece design of Figure 6.7 is very attractive, and the savings on assembly and the improvement in quality might be enough to outweigh the high costs of the material and the winding process.

Two useful guides to choice of materials are figures of merit and the relation between form and manufacturing process, and these will be studied in the next two sections.

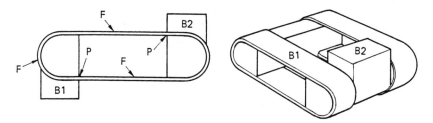

Figure 6.7 Composite spring

6.5 FIGURES OF MERIT

A figure of merit is a single parameter for a given material which is a function of its properties and its cost and will serve as a measure of its suitability for a given purpose.

As a simple example, imagine that we require a material for a tie to carry a tensile load P. Its cross-sectional area will be P/f, where f is the design stress in the given case, and its mass per unit length will be $\rho P/f$, where ρ is the density. If the cost per unit mass is c, then the cost per unit length will be $c\rho P/f$ and the cost per unit load per unit length will be $c\rho/f$. Now a figure of merit ought to be such that a high number is better, so if we invert this expression to obtain

$$f/c\rho \qquad (6.3)$$

we have the product of (force times distance) that can be sustained per unit cost, or the useful *pertinacity* (Section 2.11) per unit cost. If we take mild steel, with f at 160 N/mm², and c at £400/tonne, since ρ is about 7.8 tonne/m³:

$$\frac{f}{c\rho} = 51 \text{ kJ/£}$$

Mild steel in fact provides a cheap form of pertinacity or holding-together-power.

Figures of merit do not usually include either cost or design stress. Costs vary from day to day, unlike material properties, and also vary excessively according to who is buying and where they are, so data are both restricted in application and ephemeral.

Design stress is also an unsatisfactory parameter for use in figures of merit, because it is so variable with the application, even to the extent that it may vary from place to place in the same component. It is usual to find the ultimate tensile strength used instead, unsatisfactory as it is in its own way.

Very commonly, figures of merit are related not to cost, but to weight (in the grocer's sense: more properly, mass). Instead of the relevant unit being the pound sterling, it is the pound avoirdupois, or the kilogram. If we take the ultimate tensile strength u of mild steel as 480 N/mm², then the corresponding figure of merit is

$$\frac{u}{\rho} = 61.5 \text{ kJ/kg}$$

that is, one kilogram of mild steel would sustain a load of 61.5 N over a kilometre, and so on. Notice that this is just the specific strength.

From the argument about flywheels in Section 2.11, where it was shown that the energy stored is half the pertinacity, the same figure of merit does for a flywheel material, with the consequence, somewhat surprising at first, that low densities are to be preferred.

The best-known group of figures of merit are those which relate to the stiffness of materials, and these reflect the considerations discussed in Section 6.3, with low density being at a premium.

For example, consider a slender column used in compression to carry a load P over a length L. From equation (2.2), this demands that

$$EI = PL^2/\pi^2$$

which means that I is proportional to E^{-1}. Now I is proportional to the square of the area A for any fixed cross-sectional shape, so that A is proportional to $E^{-1/2}$. It follows that the volume of material required is proportional to $E^{-1/2}$, and the mass correspondingly to $\rho E^{-1/2}$. The appropriate figure of merit will be

$$E^{1/2}/\rho \tag{6.4}$$

Consider now a wide plate, unsupported at two edges, and loaded across the other two in compression in its own plane, the case discussed in Section 2.4. Here the stiffness EI is proportional to the cube of the thickness, and so the figure of merit becomes

$$E^{1/3}/\rho \tag{6.5}$$

It might be objected that the modulus in this expression ought properly to be E', not E (see Section 2.4), but in practice E is commonly used. The difference is small, and varies little from material to material.

Figures of merit for springs

In Section 2.10 it was shown that the energy stored per unit volume of a coil spring was given by

$$\frac{g^2}{4G}$$

where g is the design stress in shear and G is the shear modulus. It follows that the energy stored per unit mass is

$$\frac{g^2}{4G\rho}$$

so that a figure of merit for a material for coil springs is

$$\frac{g^2}{G\rho}$$

For a leaf spring, the appropriate form is $f^2/E\sigma$, or, using the ultimate tensile strength u instead of f, as explained above:

$$\frac{u^2}{E\rho} \tag{6.6}$$

Since the shear strength and modulus are generally closely related to the direct properties u and E, it is usual to use the form (6.6) as a representative figure of merit for spring materials.

A cryogenic tank

As a more interesting example of a figure of merit, consider the tank shown in Figure 6.8a, which is for carrying liquefied gas at very low temperatures by rail. The inner tank T containing the liquid is supported within an outer cylindrical casing by supports of some kind, the remainder of the space between the two vessels being filled with a non-load-bearing insulating material, either a powder or a substance like cotton wool. What material should the supports be made of? Since they have to carry loads, their cross-sectional area will be inversely proportional to f, the allowable stress. But they will conduct heat in proportion to their cross-sectional area times their thermal conductivity k. The heat leak therefore will be proportional to

$$k/f$$

The figure of merit for the material of the supports as far as heat leak is concerned, is thus

$$f/k$$

and as heat-leak is much more expensive than the supports themselves are likely to be, it is a good basis for the choice of material. By this criterion, stainless steel comes out to be a good material, because although k is large (but low for a metal) f is very large also, and the ratio is favourable. The supports will not conduct much heat, because although they are metal, they can be quite thin.

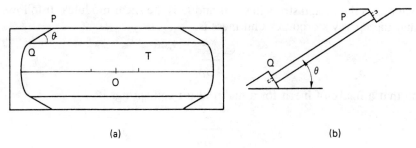

Figure 6.8 Cryogenic rail tank

The design that resulted in this case is shown in Figure 6.8a. The inner tank, of stainless steel, was supported by two sets of thin stainless steel straps, as at PQ, set at a particular angle θ, so that when cooling to the liquid gas temperature caused the inner tank to contract, the distance PQ shortened by just the amount that the strap would when the end Q was cold and the end P was at ambient. The reader may like to prove that the condition for this, if the inner tank and the straps are of the same material, is that a line from O, the centre of the tank, to the mid-point of PQ meets PQ at right angles: the method of velocity or displacement diagrams shows this very easily.

Inclining the straps in this way is essential to avoid thermal stresses. It is also good practice because the tank will be thin, and can best accept a tangential support. This same principle is seen also in the Wasa design of connecting rod which is discussed in Chapter 10. The inclination of the straps also lengthens the heat flow path, but this is partly offset by the increased area of cross-section which is needed, because the gravity load, for example, will cause greater forces in inclined members than it would in vertical ones. There is an opportunity here for making an arbitrary decision and placing the straps at equal angles round the periphery of the tank. Any such decision should be the result of reasoned argument. It might seem sensible to put more straps near the top, because of the dead load. On the other hand, such an arrangement will mean that shunting loads will be resisted more at the top than the bottom. It is a difficult matter, involving some heavy computations, to unravel these arguments.

A full study should also look at using compression supports, a scheme which would look very similar but work in a different way. Struts would be used instead of straps at positions such as PQ, as shown diagrammatically in Figure 6.8b. The angle θ would need to be changed if the material of the struts had a different coefficient of thermal expansion to that of the inner tank. Again, stainless steel would be a strong competitor, though now in a thin-walled tube to resist buckling. The small relative angular movements at the ends of the struts could be accommodated in the joints, which might be simple butt joints with a central axial pin, for register and to keep the struts in place. There are all sorts of difficult aspects, like the possibility of buckling of the inner tank wall, especially with struts, and with the strap scheme, of the rear end of the inner tank being unstable in a shunt. The straps or struts could be biased to work in both tension and compression, and so on.

Notice that the low temperatures here are an important environmental consideration, for ordinary steels and most polymers would be brittle at them.

Other figures of merit

Many figures of merit can be derived for particular problems. For example, it might be required to provide a round rod of length L and radius r, capable of accepting a twist θ while carrying a tensile load T (see Figure 6.9). The trouble is, that increasing the diameter of the rod will reduce the direct stress due to T but increase the torsional stress due to θ. The load T favours a material with a high design stress f, but the twist θ favours a low shear modulus G. If we base the allowable stress in a case of combined

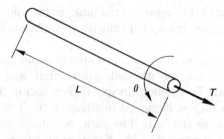

Figure 6.9 Combined load T and twist θ

direct and shear stress on the maximum shear stress (Tresca criterion), if σ is the direct stress due to T and τ the shear stress due to θ:

$$\sigma^2 + 4\tau^2 = f^2 \tag{6.7}$$

Also:

$$\sigma = \frac{T}{\pi r^2}, \quad \tau = \frac{G\theta r}{L} \tag{6.8}$$

Substituting for σ, τ from equations (6.8) in equation (6.7) gives

$$f^2 = \frac{T^2}{\pi^2 r^4} + \frac{4G^2\theta^2 r^2}{L^2} \tag{6.9}$$

Differentiating the right-hand side of this equation with respect to r^2 and putting it equal to zero gives the value of r which minimises f^2, and substitution of this r in equation (6.9) gives the minimum value of f^2 as

$$f^2 = 3\left(\frac{2G^2 T\theta^2}{\pi L^2}\right)^{2/3}$$

Thus, a design will be possible in a material if

$$\frac{f^3}{G^2} > \frac{6\sqrt{3}\, T\theta^2}{\pi L^2} \tag{6.10}$$

Thus f^3/G^2 is a figure of merit for this case, and the only materials adequate are those for which it has a sufficiently high value.

When the optimum value of r^3, which is $TL/2^{1/2}\pi G\theta$, is inserted in equation (6.9), it is found that the two terms on the right-hand side are in the ratio 1:2 or

$$\sigma = \sqrt{2}\,\tau$$

At the optimum, f^2 is shared in the ratio 1:2 between the two functions of sustaining the direct load T and twisting through θ. We have solved a disposition problem like that of the suspension bridge tower referred to in

Section 5.11. The simple ratio between the proportions of f^2 given to the two functions, of 1:2, is typical (although other simple ratios occur).

This example is unrealistic, in that the requirements could be eased by using another section rather than a circular one. The rear suspension in Section 1.8, which united the trailing arms by a member strong in bending but easily twisted, is a similar but realistic example, in which an open section is used. Other realistic examples of the same kind are the diaphragms and pins discussed in Section 8.6 as means to enable planet pinions in an epicyclic gear to be self-aligning. In each case, when the form has been optimised for the given task and the space available, we can find the combination of material properties required; if there is no available material which can meet this criterion, then we must alter something; perhaps we can reduce the severity of the task, or provide more space.

General remarks on figures of merit

Figures of merit are helpful in broad decisions on what kind of material may be suitable, as in the case of the cryogenic rail tanker. Some of the common ones have been included in the construction of computer aids to material selection, but designers may need to develop their own from time to time.

One limitation of their use is that frequently one or other of the quantities involved does not vary much. For instance, Young's modulus E does not vary significantly among steels, so in cases like springs or flexible elements, which have much in common, we are just looking for a high value of f (usually in fatigue). However, in the increasing number of cases where we can afford to look at more exotic solutions, both titanium alloys and beryllium–copper are attractive because of their lower elastic moduli, which may compensate for their lower strength.

In the case of polymers, however, there is a greater variety of properties, besides a vast number of alternative materials.

6.6 THE RELATION BETWEEN FORM AND MANUFACTURING METHOD

This is perhaps the most important influence on the choice of material, and it is certainly a very complex one. The designer should have in mind at a very early stage an idea of how every important part is going to be made, and this will limit his choice of material. Sometimes there will be two pairs kept in mind, each of a material and a method of manufacture, for example, either a sheet metal pressing or an injection moulding in a polymer. Certain groups of materials lend themselves to certain manufacturing processes, particularly when large low-performance parts, especially structures, are concerned: the lawn-mower casing is an example of this kind, as are car bodies. Some materials are difficult to form at all, for instance very hard ones.

Hard materials

A range of techniques is possible. For instance, with hard metals we can:

(a) use a cutting tool made of a very hard material, such as diamond;
(b) use a very hard abrasive, in a grinding wheel or as a slurry;
(c) use methods such as electrochemical or electrical discharge machining;
(d) use a laser;
(e) make the metal into a fine powder and then sinter it; or sometimes.
(f) cut the material to shape in a soft state, and then harden it.

As an example of the last approach, high-performance gear wheels are made from low-alloy steels, cut to shape and carburised and hardened. Unfortunately, the heat treatment in the last stage, particularly the quenching, leads to distortion, so that an allowance must be left and the final form produced by grinding. Gears of slightly lower performance may be produced by using surface hardening techniques which produce less distortion. Another interesting example of the same kind is provided by Mullite, a ceramic used for making hard insulators capable of resisting high temperatures. This is supplied as a hard clay which can be cut easily, after which it is fired to give a very hard final product.

Making the fine powder for sintering is expensive, but for small components this cost does not outweigh the savings in producing small hard components which require no machining.

Frequently great hardness is needed only locally, indeed, it may be desirable to avoid hardness elsewhere to retain toughness. This can be achieved by local surface hardening, such as case-hardening or nitriding, or by depositing by welding a local layer of a material such as stellite, or by brazing in a hard insert, as is done with machine tool cutters or masonry drills.

A cutting tool such as is used for turning may be seen as having two functions, that of the cutting edge itself and that of providing a structural support for that edge, the shank. Often now these functions are separated, and such tools consist of a reusable shank with a renewable tip with several cutting edges which can be used in turn. Each part is made of the most appropriate materials by the most appropriate process for its function. The hard but brittle cutting tip is sintered, and the steel shank, tougher but less hard, is wrought.

Decomposition of designs to suit materials and processes

Most designers engaged on a design probably have a fairly clear notion of the materials and processes to be used from quite an early stage, and think of their schemes as composed of parts. However, it is possible to imagine a design carried through without any consideration, for example, of the necessity to divide a casing into two or more parts in order to introduce the contents, gears and shafts, say. Such designs are met in nature, where the skull grows round the brain, for instance, but a designer who drew general arrangements in such a style would have to decide subsequently where to put split planes etc., to make assembly possible, which is what is meant here by 'decomposition'.

There are generally a great many ways in which a design may be decomposed, but some will be better than others. A good solution will have few joints and will help reduce the costs of manufacture, for instance, by reducing the number of setting-up operations required for the machining. Sometimes the joint between two parts may be located so that a cheaper material may be used for one of them, or perhaps a choice may be made which is better on other grounds. A classic case arose in the design of aircraft gas turbines.

Gas turbine blades and discs

Gas turbine rotor blades run very hot and must therefore be made of refractory superalloys. In the early days they were carried on discs by means of fir-tree roots located at a radius immediately inside the gas passage, where the temperature was nearly as high as at the blades (Figure 6.10a). This meant that the disc material also had to be refractory, to nearly the same extent as the blades themselves. Austenitic steels were used to provide adequate creep strength at the temperatures involved. On the other hand, where temperatures were lower, the designer would have chosen a ferritic steel which was superior in all respects other than heat resistance. The ferritic steel was stronger, cheaper, easier to machine and had a lower coefficient of thermal expansion. This last property was desirable because turbine discs, being hotter at the rim than at the hub, are subject to stress due to differential thermal expansion, the hoop stresses being compressive in the rim and tensile in the bore, and the radial stresses in the web being tensile. These useless thermal stresses play no part in resisting the centrifugal loading from the blades, but use up much of the design stress, so that effectively the austenitic disc is weaker still, and must be made even thicker.

By moving the fir-trees in to a smaller radius where the temperature was lower and connecting them to the blade by an extension, as in Figure 6.10b, it was made possible to use a ferritic disc, and savings were made of

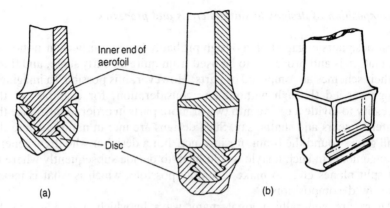

Figure 6.10 Gas turbine blade roots

up to 55 per cent on the weight of a bladed disc, important in an aircraft turbine.

The abstract view is as follows. Ideally, disc and blades would be one homogeneous structure. For practical reasons, disc and blades must be separate parts, and the question arises of at what radius the junction should be. This is a standard problem in decomposition. The answer is, at a radius just small enough to be cool enough to make a ferritic disc practicable.

Similar thinking has been behind some steam turbine casing design, with an inner shell of more refractory metal surrounded by a cooler and less refractory outer casing, the space between being full of steam at an intermediate pressure and temperature.

Separation of function to suit materials

Sometimes, as in the gas turbine case just discussed, the 'monolithic' design is decomposed into manufacturable pieces in such a way as to suit a preferred choice of materials, but temperature is rarely the basis. Usually, separation of function (Chapter 1) is involved. A common case is when a relatively small insert of one material is made in another, for example, a brass bush in an iron casting to serve as a bearing, or a threaded brass insert in a polymer moulding to provide for attachment by a metal screw. The case of the large pressure vessels of advanced gas-cooled reactors has already been referred to in Chapter 1, and a further example is given in Chapter 7, a piston and rack made mostly of polymer, but with the rack teeth in stainless steel. This last case shows the familiar pattern of a large quantity of a cheap material combined with a small quantity of an expensive one where the duty requires special properties, here, greater strength. In the case of the brass bush, the special property is not extra strength, but that of forming a better bearing pair with a steel or cast-iron journal.

The moulding of forms stiff in torsion

It was seen in Section 2.5 that open sections are weak and easily twisted in torsion, in contrast to closed ones. A product like the bed of a lathe in cast iron, or the shaft of a toy spade in polymer, may be required to be fairly stiff against twisting in use, but it is difficult to cast or mould a closed section. This problem can be overcome by the form shown in Figure 6.11: this may be several times less stiff than a closed section of similar scantlings, but it is very much better than an unbraced open section and yet it is still suitable for moulding, since it can be made in a two-part die, or cast without needing cores.

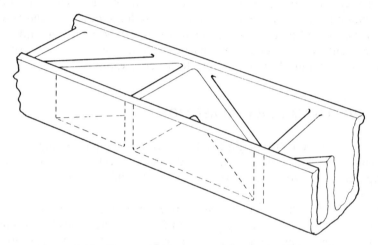

Figure 6.11 Form suitable for moulding and stiff in torsion

There are many other instances which could be quoted of design to suit materials and manufacturing methods. Great ingenuity has been displayed in using pressings, for example, and overcoming the limitations of this otherwise very attractive replicative process. Some of these trends have been helped by the development of new materials and processes.

6.7 NEW MATERIALS AND PROCESSES

New developments in materials and processes are reported every week in the engineering press, and the designer has a hard task keeping reasonably abreast of progress. The most dramatic developments in materials are perhaps the light, high-strength composites with resin matrixes reinforced with carbon, boron or polyaramid fibres, or the turbine blades made from a single crystal, or ceramic-fibre reinforced metals. Then there is the slow but steady growth of the use of ceramics, for bearings, for instance, but soon perhaps for moderately stressed hot engine components. With these glamorous developments, it is easy to overlook the vast strides in casting, especially of aluminium alloys, in the use of adhesives, in powder metallurgy and in the properties and processing of polymers.

To take just one example, the choice for casting aluminium alloys used to be between die-casting, with relatively tight tolerances and a superior finish, and sand-casting, with lower costs and greater freedom of form because cores could be used. In recent years, however, the quality of sand-casting has been improved markedly, so the designer has had to change his thinking.

Even social considerations must influence the designer: aluminium alloys seemed at one time an excellent material for railings for footbridges and similar places, but recently theft of such railings for their value as scrap metal has caused a reassessment of this view. A possible contender as a replacement is glass-reinforced plastic sections formed by 'pultrusion', expensive in themselves but of very little scrap value.

6.8 AIDS TO MATERIAL SELECTION

In recent years much thought has been given to helping designers in the difficult work of material selection, and much of this has been embodied in computer aids, with the power of calling-up data as required from a database.

As a first step, however, the charts produced by M. F. Ashby are very helpful. He has kindly given permission for one to be reproduced (Figure 6.12). It shows Young's modulus E plotted against density ρ on log/log scales for most useful materials. The engineering metal alloys are shown in the top right-hand corner, with a large, rather triangular, bubble enclosing them all and a number of small bubbles within representing particular groups. Thus steels are represented by a bubble near the middle of the top side of the large bubble; the small size of this bubble reflects the small range of variation of E and ρ in steels.

The group of four parallel broken guide lines are lines of constant specific stiffness E/ρ, and the two other broken lines correspond to constant values of the figures of merit in expressions (6.4) and (6.5).

Notice the two bubbles for the woods, corresponding to stiffness along and across the grain, but with nearly constant specific stiffness, E/ρ, along the grain, and the polymers, with a relatively small range of density but a huge range of stiffness. The highest specific stiffness is that of diamond, followed a long way behind by other engineering ceramics and then carbon-fibre reinforced polymers in the unidirectional form. Steel has only about one-tenth the specific stiffness of diamond, but usually it is the stiffest material the designer can command.

MATERIALS AND MANUFACTURING METHODS

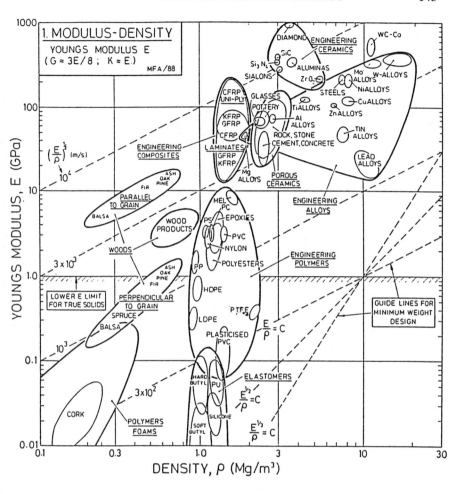

Figure 6.12 Modulus–density

The other charts in the series cover other combinations of properties, including many which have not been discussed here.

Notwithstanding the relation of material to function, manufacture is nearly always an important constraint on choice, and the designer must consider material, manufacture and function all at once, as the next chapter shows.

7

Pneumatic Quarter-turn Actuators

7.1 INTRODUCTION

A good example of design is provided by pneumatic quarter-turn actuators. These devices use compressed air at a gauge pressure of up to about 7 bar to turn valves through 90°: the valves in question are usually ball valves in which a sphere with a hole in it can be rotated about a diameter so that it allows the free passage of fluid or completely blocks the flow. The actuators come in many different kinds and sizes and are extensively used in the chemical and process industries. Much ingenuity has been displayed in designing them for low cost and high reliability.

Figure 7.1 shows the working principle of several types of these actuators. Those shown in Figure 7.1a and b are perhaps the most common, both using pistons which convert their linear motion to rotation via racks and pinions: the first has two double-acting pistons and the second has two single-acting pistons joined by a common piston rod. Figure 7.1c shows a vane type, which has an appealing simplicity, and Figure 7.1d a type in which a pair of single-acting pistons united by a single piston rod acts on a slider working in slots in a crank. This last design may appear clumsy, but has one distinct advantage over the others, in that it is *well-matched* to its task. The torque it develops is greater at the ends of the quarter-turn than in the middle, and it is at the ends that the resistance to turning is greatest: the other three all develop a uniform torque, and so require a larger swept volume, as we shall see.

The advantages and disadvantages of these and other *configurations* present a complex network of considerations. For example, designs like Figure 7.1a exert a pure couple on the output shaft, compared with the large unbalanced loads in the other three, so that the bearings for the

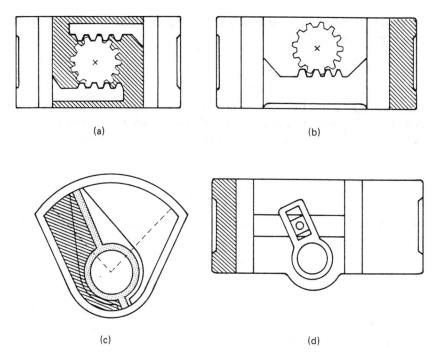

Figure 7.1 Configurations of quarter-turn actuator. X indicates the output shaft. The shading indicates the space pressurised for clockwise rotation

output shaft need be no more than simple bores in an aluminium alloy casing: the others require proper bushes. In the arrangements of Figure 7.1b and d, the central space containing the output shaft is not pressurised, so that sealing is less demanding and there is more freedom in the decomposition problem, the dividing up of the structure into parts which can be manufactured and assembled.

Some arrangements lend themselves better to particular materials and manufacturing processes. For instance, Figure 7.1b can readily use stainless steel pressings for the cylinders whereas Figure 7.1a cannot (or at least, it seems to the writer that it cannot). Indeed, this is one of the many design problems in which the relationship between configuration, materials and manufacturing process should be prominent from the early stages.

7.2 FUNDAMENTAL CONSIDERATIONS

The problem field of the quarter-turn actuator can be mapped out in a few essential respects by studying fundamental mechanical aspects. If the swept volume, the volume of compressed air admitted on one 'stroke', is V, and the gauge pressure is p, then the work done by the air admitted is pV. If the torque generated is Q then

$$\int_{-\pi/4 - \epsilon/2}^{\pi/4 + \epsilon/2} Q d\theta \tag{7.1}$$

is the work done on the valve, where θ is the rotation and ϵ is the small extra travel beyond the quarter-turn which is provided to cover tolerances. In most cases (but not in Figure 7.1d), Q is constant so that expression (7.1) reduces to

$$Q\left(\frac{\pi}{2} + \epsilon\right)$$

and if there are no losses, this product must be equal to the work done by the air entering, or

$$Q\left(\frac{\pi}{2} + \epsilon\right) = pV \tag{7.2}$$

In a given specification, Q, p and ϵ will be laid down, so that V can be calculated. However, in real cases there will be friction losses, so that we need to introduce an efficiency η, and

$$Q\left(\frac{\pi}{2} + \epsilon\right) = \eta p V \tag{7.3}$$

As an example, suppose the actuator must be able to deliver a torque of 100 Nm when ϵ is 0.05 radians and p is 0.5 MN/m². A good design will normally have an efficiency of about 0.8, so we can estimate V as

$$V = \frac{Q\left(\dfrac{\pi}{2} + \epsilon\right)}{\eta p} = \frac{100(1.57 + 0.05)}{0.8 \times 0.5 \times 10^6} \, m^3 = 405 \, cc$$

(While the cubic centimetre is not an SI unit, it is the most convenient here and still widely used and understood).

Compactness

Suppose the actuator is of the piston, rack and pinion type, for instance, like those in Figure 7.1a and b. and has n pistons of area A operating on each stroke. Then if the pitch radius of the pinion is r, the torque without friction would be $npAr$, and with an efficiency η:

$$Q = \eta n p A r \tag{7.4}$$

The stroke of the pistons is $(\pi/2 + \epsilon)r$, so the swept volume is $(\pi/2 + \epsilon)nAr = V$, which shows the relation between equations (7.3) and (7.4).

From equation (7.3), the swept volume is fixed by the torque, the working pressure and the efficiency. As the cost of the actuator is likely to be closely related to its overall size, then, other things being equal, a good design will be one for which V is large compared with the overall volume. Compare the schemes in Figures 7.1a and b: in the first, the pistons are double-acting and so n in equation (7.4) is 2. If air is admitted to the central space, both pistons are working to turn the shaft clockwise. In Figure 7.1b, air is admitted to the right-hand end only when turning clockwise, and acts only on one piston, so n is 1, and this design with the same piston diameter and stroke will produce only about half as much torque. To produce the same torque, Figure 7.1b needs to be about 25 per cent bigger each way $(1.25^3 \simeq 2)$. A little thought will show that Figure 7.1c is very compact, while Figure 7.1d is large, but not so large as Figure 7.1b, because of the good *matching* noted earlier.

Consider the forces acting in the crank and slider scheme of Figure 7.1d, but neglect friction. If F is the force the air exerts on the piston, this is balanced by two reactions, one from the cylinder walls at right angles to the axis and one from the slot or guides in the crank which must be at right angles to the slot. If θ is the angle between the crank and its central position, these reactions must be $F\tan\theta$ and $F\sec\theta$, as shown in Figure 7.2.

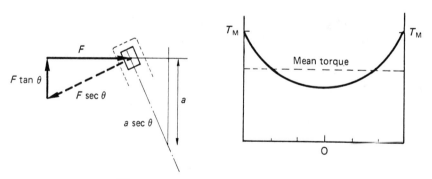

Figure 7.2 Force analysis, Figure 7.1d

If a is the distance between the cylinder axis and the axis of the output shaft, then the force $F\sec\theta$ acts on the crank at a radius $a\sec\theta$, producing a torque $Fa\sec^2\theta$, which varies as shown in the graph at the right of Figure 7.2. The maximum torque, at the ends of the stroke where θ is $\pm\pi/4$, is $2Fa$, and that in the middle is only Fa. Integration of $\sec^2\theta$ shows that the average torque is $4Fa/\pi$. Now the shape of the graph matches the task well, giving the maximum torque at the ends of the stroke, which is where the valve is hardest to turn (the break-out torque). The effect of this is that the mean torque is only $2/\pi$ times the maximum torque which determines the rating, and so the swept volume need only be 0.64 of what it would be for Figure 7.1b, say. As far as overall size goes, the schemes rank in the order c, a, d, b, but this is offset in the case of c by the awkward shape.

7.3 ASPECT RATIO

The preceding arguments give values for the swept volume and for Ar, the product of piston area and pinion radius, leaving open the choice of A or r. Should we choose a small A and a large r, or vice-versa? A crude but sound approach suggests that neither a small A, giving a large r and a long thin actuator, nor a large A, giving a small r and a short fat actuator, is desirable. Both will have the same interior volume as a middling design to give the same torque, but both will have a relatively large surface and hence a large amount of casing. Memories that a sphere has the minimum surface area for a fixed volume may suggest that the stroke should be about equal to the radius of the pistons, and that is about right.

There is, however, another consideration which may lead us away from the fat design, which is the stresses in the rack teeth: this depends, however, again on the materials and methods of manufacture. Let us see first the effect of proportion, and then relate that effect to materials. The technique adopted is one of great simplicity and also great value.

Suppose we have a design of actuator which has a cylinder radius R and stress f in the gear teeth. Keeping the torque Q the same, imagine R increased to kR ($k > 1$). Then the force F on a piston increases to k^2F. Because the torque is the same, Ar remains the same, and since A has been increased by a factor k^2, r becomes r/k^2. The strength of the teeth per unit width is proportional to their height, and so is reduced by a factor $1/k^2$, but the face width can be greater in the ratio k, so the strength overall of the teeth is multiplied by $k/k^2 = 1/k$. The load F on the teeth has been increased by a factor k^2, and their strength has been decreased by a factor $1/k$, so f, the stress in the teeth, has become k^3f. To summarise, as the pistons are enlarged, the stroke and hence the pinion radius r become smaller, and the teeth become weaker even if the face width is increased as much as possible. The load on the teeth is greatly increased, so the stress on them increases rapidly.

At some aspect ratio (the ratio of the piston diameter to the stroke), this effect will become limiting. The actual value of the ratio at which this happens will depend on the material of the teeth. We shall return to this later.

7.4 TABLE OF OPTIONS

In Chapter 1, combinative methods of conceptual design and tables of options were discussed. These aids belong to the first, or conceptual, stage of design but their use can be extended into the embodiment stage, and the following discussion will cross that boundary, which is a vague one at best.

One of the difficulties of the combinative technique is that it is a Procrustean device (Procrustes was a legendary Greek robber who had a bed to which he made all his guests fit, either stretching them to suit or cutting off their extremities). Usually not all the possible designs will fit conveniently into the same table. Thus, one function in a table of options for the quarter-turn actuator might be 'convert linear to rotary motion', a function which does not exist in the vane type shown in Figure 7.1c. A single anomalous aspect like this does not matter much, but when there are many, as may well happen, the situation can be very untidy.

However, when a few basic decisions have been taken such problems disappear. Table 7.1 shows a table of options for an actuator of the rack-and-pinion kind, with two double-acting pistons, like the one in Figure 7.1c. It has only four functions or lines, because it is a *kernel* table, giving only those functions which bear most strongly on the design at the stage reached. It is only after some study of the problem that the choice of the kernel functions can be made, and even then it may prove to be a poor choice later.

The late stage kernel table (Table 7.1) contains only those functions which have a strong influence on the design after the decision has been

Table 7.1 Table of options, forms in Figure 7.1d

Function	Means
A Contain pressure	1 Aluminium alloy extrusion etc. 2 Aluminium alloy castings 3 Stainless steel pressings 4 Reinforced polymer mouldings
B Pistons and racks	1 Aluminium alloy castings 2 Polymer pistons with steel racks
C Guide pistons	1 Long skirt on pistons 2 Guide rods fixed in ends 3 Guide rods fixed in pistons
D Distribute air	1 Hollow guide rods 2 Passages in casings 3 Transfer slots 4 Hollow piston rods

made to have two opposed double-acting pistons. They are largely concerned with materials and manufacture, which exert a major influence at this stage. For instance, A, 'contain pressure', it has already been decided to do with cylinders, but it remains to decide the material and method of manufacture. It is attractive on cost grounds to use an aluminium alloy extrusion for the body, capped with two cast ends. However, the limitation to one section throughout its length is a serious one: a casting enables more material to be provided locally, which can be a great advantage. Similarly, it would be attractive to use stainless steel pressings or cans, to provide cylinders and end caps combined, as in Figure 7.3, but this leads to difficulties with function D, distributing air to the cylinders. Passages cannot be made in the walls of the can and to braze pipes to the outside would not be acceptable. This leaves the only way of supplying air to the space in a can as through the piston, either via a passage in the piston rod, or via a fixed tube passing through the piston.

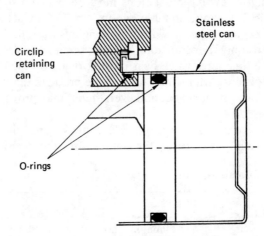

Figure 7.3 Use of 'can' as a cylinder

Studying a combination

Figure 7.4 is a rough sketch for a design with stainless steel cans, which uses fixed tubes B sliding in holes in the pistons to transfer air to the cans. The arrangement is clumsy, but can show us how to do better; such sketches are a characteristic way of working of the designer. He draws something, finds the weaknesses, sees ways round them, and proceeds a step or two further as a result. In the ideal case, a number of avenues are explored in this fashion, each until it seems that progress has ceased or become too slow. The best of all the end points is then chosen.

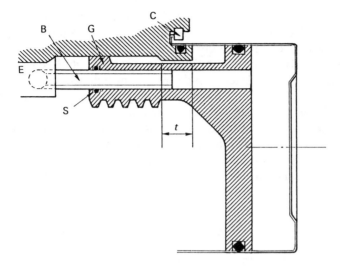

Figure 7.4 Sketch for design with stainless steel 'cans'

In Figure 7.4, the attachment of the can to the body, ultimately by the circlip C, takes up a great deal of space radially. It is undesirable to use a circlip to carry such a large load, and questionable practice to use a circlip in an aluminium alloy groove. Another undesirable feature is the block E, because it has to fit between the end of the piston rod and the underside of the other piston when the pistons are nearest together. This means that the length of E must be added to the overall length of the actuator, making it a very expensive detail. These aspects will be returned to later when the survey of the combinative framework has been completed.

The function B, pistons and racks, can be treated as one function or separated. If the piston and rack are to be made in one piece, then the material needs to be at least as strong as aluminium alloy because of the heavy loading of the rack teeth. If the rack is made a separate part and of strong material, however, it can be fixed into an injection-moulded polymer piston and rod. Because the rack is a small part, especially if it is of very strong material, it can be made by sintering sufficiently accurately to require no machining subsequently.

The guidance problem

Function C needs consideration. Figure 7.1a is quite unworkable as shown because on the in-stroke the pistons would tilt in the bore, pushing the racks away from the pinion. On the out-stroke, the teeth would be forced into contact, increasing friction. It is essential to prevent the first and desirable to avoid the second, by providing proper guides for the pistons.

The actuator provides an example of the sticky drawer problem that was discussed in Section 5.8. In the case of the piston shown in Figure 7.4, jamming would certainly occur on the in-stroke if a guide were not provided at G. But just to avoid jamming is not enough. Even with a deep guide, most of the inefficiency in designs of this kind will arise from guide friction because of the large moment resulting from the offset between the force on the piston and the balancing reaction at the teeth (see Figure 7.5). Notice that this problem scarcely exists with the actuator of the type with two single-acting pistons united by a single piston rod (Figure 7.1b), as this has a very long guide and a pinion which can be offset to reduce the moment problem (and also reduce at the same time the bending moment in the piston).

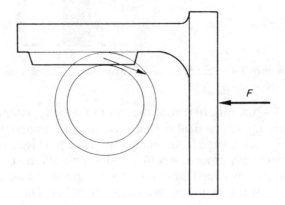

Figure 7.5   Offset load on piston

7.5 STUDYING A COMBINATION

We have surveyed the kernel table sufficiently to start looking for promising combinations. First, let us follow through one particular choice of means. The use of stainless steel cans for containing pressure (A3) was seen to be attractive but led to a characteristic problem of disposition (see Chapter 1) which was illustrated in Figure 7.4: the block E adds to the functions to be accommodated in the overall length of the machine, and so makes it longer. Is there any way of avoiding this problem?

To answer that question, consider whether it is possible to apply one of the basic resources in disposition problems, that of overlapping two of the functions which have to be fitted into the length available. A useful exercise is to look at the functions which must be squeezed into the length, expressing them all in terms of the pinion pitch circle radius r.

If the pinion has 16 teeth of module $r/8$, then its tip diameter will be $2(r + 0.125r) = 2.25r$. The stroke s, as we have seen, will be

$$\left(\frac{\pi}{2} + \epsilon\right)r$$

and if ϵ is the tolerance on angle, about 3°:

$$s \simeq 1.62r$$

so that when the pistons are at their furthest out the distance between their inner sides must be

$$(2.25 + 2 \times 1.62)r = 5.5r \text{ (say)}$$

(see Figure 7.6). If in the innermost position, each piston almost touches the pinion, then the distance QE in the figure is $1.125r$, and this is quite enough to allow plenty of overlap for the tooth engagement (a good back-up is 4 modules, or with a 16-tooth pinion of pitch radius r, $0.5r$). Since we only need $0.5r$ of the $1.125r$ available for the run-out of the rack, there is

$$(1.125 - 0.5)r = 0.625r$$

available for the block E of Figure 7.4. This is rather on the mean side, as in a small actuator r may be only 10 mm, but even if we go to $0.8r$ for E it only means lengthening the whole by

$$2 \times (0.8 - 0.625)r = 0.35r$$

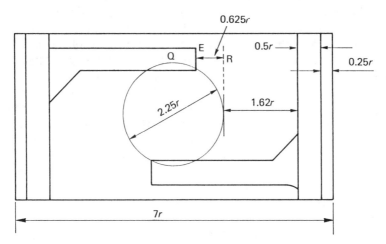

Figure 7.6 Axial disposition problem, internal

If we add in the two piston thicknesses, at say, $0.5r$ each, and $0.25r$ for each end cap, we have an overall length of $7r$, or $7.35r$ if we add $0.35r$ for the block E.

If we were to abandon symmetry we could make space for the block E on one side, leaving the other side short, giving an overall length of $7.175r$. However, this would mean making the pistons and racks different parts for each end, and the asymmetry might reduce the visual appeal, especially as it would be small and so more likely to look wrong than a large difference.

The systematic approach is to consider all the functions which are stacked up in Figure 7.6 and see if they must all lie end-to-end, the end caps, the piston thicknesses, the strokes, the pinion and the block E. Only one can overlap the others, and that is the block, which, if there was space, could be put *alongside* one of the racks instead of end on to it. There will be space if the rack is narrow, which will certainly be the case if the piston and rod are of polymer, with a sintered steel insert for the rack.

Unfortunately, there is also an axial disposition problem on the outside of the actuator. The distance between the undersides of the pistons is $2.25r$ when they are closest together (Figure 7.6) and we can see from Figure 7.4 that this length has to accommodate the pinion shaft and its bearing in the casings, and the two can-to-body seals (dimension t in Figure 7.4). But if the body is not to be split axially and the pinion is not to be separate from the shaft, the shaft diameter, on one side at least of the pinion, needs to be at least $2.25r$ to make assembly possible.

Splitting the body axially would lead to unacceptable sealing problems. A separate pinion would be possible, but because of the large torque, serrations would be needed to take the drive from pinion to shaft, increasing the cost. Even with a separate pinion, it is doubtful whether we could do better than $1.5r$ for the shaft and its journals and $0.6r$ for t, giving a total distance between the insides of the pistons at their closest of $2.7r$, against the $2.25r$ with an overall length of $7r$ for the whole actuator.

We need also to look at the same problem in a plane transverse to the shaft, where from Figure 7.4 it appears we have to dispose two seals of length t, the stroke of $1.62r$ and the block G in the minimum distance between the inside faces of the pistons. But Figure 7.4 contains an arbitrary decision, in that the block G slides on a surface recessed below the two can/body seals. By raising this surface so it runs across the seals, this particular disposition problem ceases to be critical.

The outcome of studying this combination or scheme, with two double-acting pistons and pressed stainless steel cans, has been to solve one disposition problem, concerning the block E in Figure 7.4, only to discover another, on the outside parts.

There is another difficulty, which is that of securing the cans to the body. In Figure 7.4, this was done by circlips C, but this is not a satisfactory solution, and tends to built up to a large diameter. With a large length and a large diameter, the whole scheme is unattractive.

At this point it seems wise to drop this combination altogether, but with one proviso. The *internal* axial disposition problem, involving the block E, could be solved. The *external* axial disposition problems hinges on the space required in the middle for the shaft and its bearings. This space is only required round two parts of the circumference, near the shaft axis. If the attachment of the can to the body is confined to the area between, the axial lengths needed to accommodate the attachment and the shaft overlap, and the disposition problem is slightly eased. The general form of such a can is shown in Figure 7.7, with the location of the can/body seal. The attachment would be confined to the 'ears' at A.

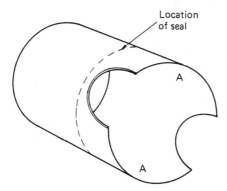

Figure 7.7   Can with ears

The longitudinal stresses in such a can are low, because the thickness needed in the ends and for general robustness in use must be much greater than simple membrane tensions dictate. For instance, with a can of diameter 60 mm and 0.8 mm thick, at a pressure of 0.7 N/mm² gauge, the axial stress would be 13 N/mm². With a structure as deep as the side wall of the can, the concentration of this tension into ears constituting two-thirds of the circumference would still leave the stresses low. However, the writer has not been able to devise a satisfying solution to the attachment problem. Folding lugs round the ends of the ears back on the main body to obtain a shallow engagement bayonet catch is not pleasing (Figure 7.8), though cheap and strong enough, and leaves the problem of manufacturing the mating part. A rolled end to the ears clamped between two arcuate clamps would answer (Figure 7.9), but looks clumsy, particularly as any through bolts would need to pass on either side of the shaft, making the axial extent of the clamp very long: also, air connections would have to pass through the clamps to the body.

156 FORM, STRUCTURE AND MECHANISM

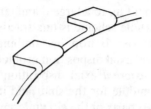

Figure 7.8 Bent-back lugs

At about this point, as there seems no end to the difficulties, it is appropriate to drop this combination and look at something more fruitful.

7.6 TWO GOOD DESIGNS

An early design of quarter-turn actuator which has proved extremely successful was the Norbro design of 1968, shown in Figure 7.9. In terms of the table of options (Table 7.1), it is described as 1, 1, 3, 1, that is, the body is an aluminium alloy extrusion, the end caps are aluminium alloy castings, the pistons and racks are combined and are also aluminium alloy castings,

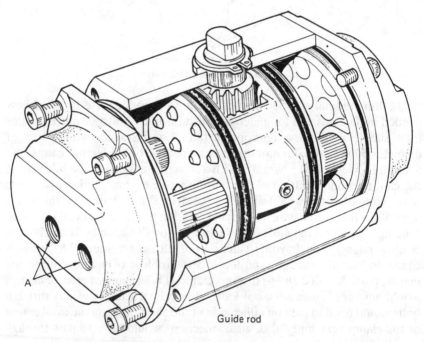

Figure 7.9 A good design of actuator (Norbro, 1968). The stainless steel guide rods act also as air distributors

PNEUMATIC QUARTER-TURN ACTUATORS

and the problem of guidance is overcome with two hollow stainless steel guide rods that are fixed one in each piston, and slide in the end caps. The last function in the table, that of distributing air, is achieved by using the hollow guide rods as air passages. The holes in which they slide in one of the end caps have air connections to them from the outside (A in Figure 7.9).

O-ring seals are provided wherever necessary, and also polymer bearing bushes where guide rods slide in end caps or in the piston to which they are not fixed. Each piston-cum-guide-rod assembly is thus supported by polymer bushes and has no direct contact with any metal except at the tooth meshing between rack and pinion.

This very elegant design has stood the test of time, though some minor improvements have been made over the years. The decision to fix the piston rods in the pistons, rather than the end caps, means that the holes in

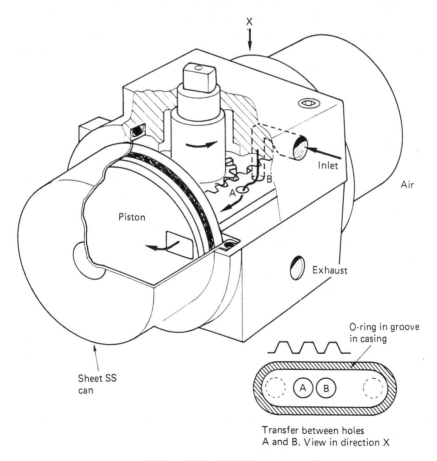

Figure 7.10 Miti-1 actuator

which they run in the end caps must be long enough to accommodate the full stroke, of $1.62r$, an O-ring seal and a bush, say, about $2.2r$ in all. A simple end-cap might have a thickness of only $0.5r$, so fixing the guide rods in the pistons instead of the end caps increases the length overall by something approaching $4r$. The extra length, however, can be turned to advantage, and there is also an advantage in the very long guidance. The air pressure on the piston and the reaction from the pinion teeth apply a moment, and by opposing this by forces from guides as far part as possible, the transverse reactions and the friction that goes with them are kept small.

In many applications, valves have to 'fail safe', reverting to a particular position if there is an interruption of the air supply. This is done by energy stored at the actuator, either in steel springs or a small compressed-air tank. In the Norbro design, coil springs are placed between one piston and the end cap, so that if the air supply fails, that piston will be forced in by the springs, turning the valve to the required position. The springs must be fairly long even when compressed, and this length is accommodated in individual holes or pockets in the end cap. In this way good use is made of the extra depth of cap which stems from fixing the guide rods in the pistons. In the second example, a different approach is taken to this spring return provision.

A *single-acting design*

The difficulties which arose in the study of the combination with stainless steel cans were largely associated with the need to seal the centre section. With a single-acting design like that in Figure 7.1b, the centre section is never pressurised and so need not be sealed. The guidance problem is solved, since each piston acts as a guide for the other, via the single piston-rod which joins them. The disadvantages of the single-acting version are its greater bulk (discussed in Section 7.2) and the fact that all the load comes on one side of the pinion, which means there are heavy bearing loads on the shaft, for which better provision must be made than a simple hole in an aluminium body. An elegant single-acting design is shown in Figure 7.10.

The body is formed of a pair of aluminium alloy die-castings which close together over two stainless steel cans with flared rims. These rims engage in grooves in the castings which secure the cans in place, the whole being held together by the bolts through the body.

The two pistons and the rod which joins them form a single polymer moulding, into which is secured the rack, sintered from stainless steel. The pinion is integral with the steel shaft and runs in generously-proportioned polymer bushes.

One function remains to provide for, the distribution of air. This is done via passages in the piston-rod, one from each end, which communicate via holes with air passages in the body. The transfer of air from body to rod is complicated by the fact that the rod slides longitudinally in the body, so that a slot must be provided so that, throughout the stroke, the air hole is opposite some part of the slot. Sealing is achieved by an O-ring in a groove surrounding the slot.

This actuator is the Worcester Controls' Miti-1: the name is as excruciating as the design is elegant.

7.7 SUMMARY

It would be possible to study some of the other designs, those on the market and some no longer available. The types shown in Figure 7.1c and d could be discussed further, and there are others, including a version with four cylinders. Enough has been said, however, to show the complicated considerations which arise even in a simple product like these. In particular, it is manufacture that often proves crucial in the long run.

8

Epicyclic Gears

8.1 THE FUNCTION OF THE PLANET CARRIER

Figure 8.1 shows the principle of an epicyclic gear. The input gear or *sun pinion* drives a number of planet gears P (commonly, three). These planet gears are mounted in a rotating planet carrier and mesh with a fixed internal or *annulus* gear round the outside of the planet carrier.

Figure 8.1 also shows the loads on a planet gear. If F is the tangential load on the teeth at the inside, exerted by the sun pinion, then this is balanced by an equal force F exerted by the annulus gear on the teeth at the outside (to see this, consider moments about the axis of the planet gear). The tooth loads are therefore equivalent to a force of $2F$ through the centre of the planet gear and this force produces reactions of F at each bearing. It was shown in Section 4.4 how the loads F may be balanced between three planets by floating either the annulus or the sun pinion transversely, an example of kinematic design.

The planet carrier is formed of two discs, each supporting the three planet bearings at one end, connected to each other by three columns placed in the spaces between the planets. Each end disc is subjected to three tangential forces F at the pitch radius a of the planet axes, giving a total output torque of $6Fa$, applied half to each disc.

One disc, D (see figure), is attached to the output shaft and the torque applied to that disc is transmitted through its plane to the shaft. However, the torque applied to the other disc C must be transmitted to disc D via the columns; this would require very little material if it could be disposed in the best way, in the form of a thin-walled tube. However, because the material can only be put in certain places, and because the structure must be very stiff, the problem is not a simple one.

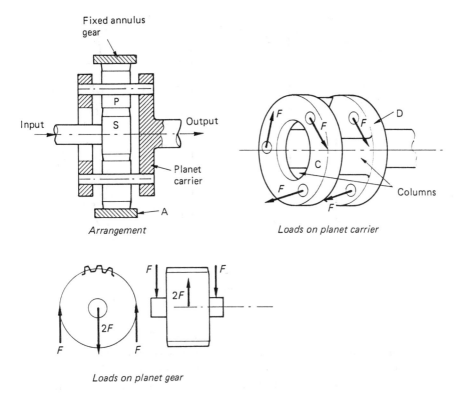

Figure 8.1 Loads in epicyclic gear

Visualising the structural behaviour

The way the structure of the carrier transmits the torque of $3Fa$ from disc C to disc D is difficult to visualise, and two viewpoints will be adopted to clarify it, the first being shown in Figure 8.2. The first of these four sketches, (a), shows the development of a simple cylinder joining the rims of discs C and D, that is, the cylinder is slit along one generator and unrolled to a flat rectangle. Because of the nature of the structural task, which is axisymmetric and fundamentally one of transmitting shear, we can adopt this simplified view without error. The ideal thin tube structure is not possible, of course, because it needs to pass through the planets: this is another intersection problem. One planet is shown in broken lines in Figure 8.2b, which also shows the six forces F applied along the ends of the tube, which correspond to the rims of discs C and D, and the shearing of the rectangle which represents the tube. This shear is vastly exaggerated for clarity: in a typical case it would be of the order of one-thousandth of a degree. The top and bottom edges of the rectangle are actually joined, and

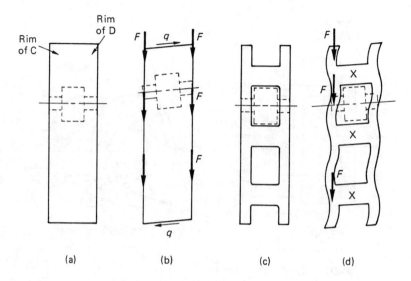

Figure 8.2 Developments of the carrier

there will be a shear flow along them, shown as q. The material of the tube is all working in pure shear, and the structure is very nearly an ideal one: the ideal or Michell structure would be a tube which would belly outwards and increase slightly in wall thickness between C and D, and would save one or two per cent of material.

It will be seen that the effect of deflection is to turn the planets at an angle to the axis of the gear box, so that they mesh badly with the sun and the annulus, concentrating the loads on the ends of the teeth towards the output shaft on both inner and outer meshes. This concentration of load at the D end reduces the torque in the carrier. Nevertheless the structure needs to be very stiff for this effect to be acceptable. The stresses in such a planet carrier are low, because so much material is needed to give the necessary stiffness: it is a *stiffness-determined* structure.

Returning to the intersection problem, we can only use the space around the planets for the carrier structure. Figure 8.2c shows such a structure in its essence, a tube (shown developed) with three rectangular holes in way of the three planets. This will deflect after the fashion shown in Figure 8.2d, largely by bending. With the proportions shown, this needs about forty times as much material for the same stiffness as the simple tube: this is the essence of the problem.

The aircraft structural engineer and others would call the holes in the structure of Figure 8.2c 'cut-outs', a good, plain, descriptive term. The cut-outs increase the stresses a great deal and the flexibility much more. Let us look at the ways we can minimise the flexibility by good disposition

8.2 INCREASING THE STIFFNESS BY ASYMMETRY

To understand how the stiffness can be increased it is necessary to have a clear picture of how the structure works, and to this end use will be made of a more insightful representation even than the developments of Figure 8.2, helpful as they are. From the deflected form in Figure 8.2d it can be seen that there is a *contraflexure* in the columns (at X), where the curvature, and hence also the bending moment, change sign. At these points a hinge could be inserted which would not affect the structural behaviour.

The carrier is equivalent structurally to two three-legged stools with their legs joined together at the point of contraflexure, X, as in Figure 8.3a. Each stool is subjected to a torque $3Fa$, balanced by a torque $6Fa$ on stool D, to which the output shaft is attached. If we regard the jointing of the legs as by a hinge or ball joint, something not capable of transmitting moment but only shear, then the junctions of the legs will correspond to the points X of zero bending moment in Figure 8.2d. The shear at each of the points X is F, as in Figure 8.3b, which shows the C stool taken in isolation to be in equilibrium, as of course it must be.

Now this 'stool' picture of the structural functioning of the carrier is lucid and insightful, and will enable us to improve the design. Most of the deflection is likely to be due to the bending in the 'seat', the disc C, which will bend into a wavy form of three wavelengths, as shown by the broken line, under the action of the three moments FL applied by the legs, where L is the leg length. The slope of the wave at the top of the leg will be

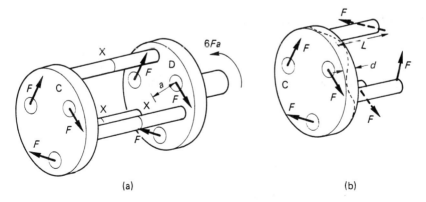

Figure 8.3 Two-stool equivalent structure

proportional to these moments and inversely proportional to the bending stiffness of the seat or disc, which is itself proportional to the thickness of the disc cubed, from bending theory (remember $I = bd^3/12$, Chapter 2). If we ignore bending in the columns, the slopes of both discs will be the same, and so X will be positioned in the column such that the L's are proportional to the bending stiffnesses since

$$\text{slope is proportional to} \frac{\text{moment}}{\text{bending stiffness}}$$

Now consider two extremes of the design possibilities, one with discs of equal thickness d and one with one disc of thickness $2d$ and one very thin disc, so that both designs use roughly the same amount of material. Because of its very small bending stiffness, L for the thin disc in the asymmetrical design will be very small, so L for the disc $2d$ thick will be virtually the whole length of the columns, twice as great as in the symmetrical design. However, the stiffness for the thick disc will be 2^3 times as great, so the slope will be $2/2^3 = 0.25$ times as great, that is, the carrier will be four times as stiff, and the slope on the planet pinions will be reduced to a quarter.

The problem is much more complicated than this simple analysis suggests. For instance, the thin disc cannot be of zero stiffness and use no material, so the full four times improvement cannot be obtained. Moreover, discs of uniform thickness will not use material as efficiently in bending as discs with ribs, so the bending argument must be treated in more depth. Finally, there is a little bending in the columns which alters our results slightly. Nevertheless, the basic finding is sound, that an asymmetrical disposition will be much more economical in terms of stiffness per unit mass.

8.3 LOCAL FORM DESIGN

Figure 8.4a shows the end view of the thick disc in a more developed form, where the circles B represent the bearing housings for the planets. Instead of a solid slab a thin disc has been provided with two stiffening ribs G, so that the section AA is a channel form. This is a more economical way of resisting bending, as was seen in Chapter 2. However, there is a weakness in this structural design in that the curved ribs G will not work as well in bending as straight ones would.

In Figure 8.4a, the radial broken lines represent the columns C on the far side of the disc. The circles B are the bearing housings. Consider the length of the outer rib G between two columns, HK in the figure. If the torque on the disc from the forces F on the other ends of the columns is

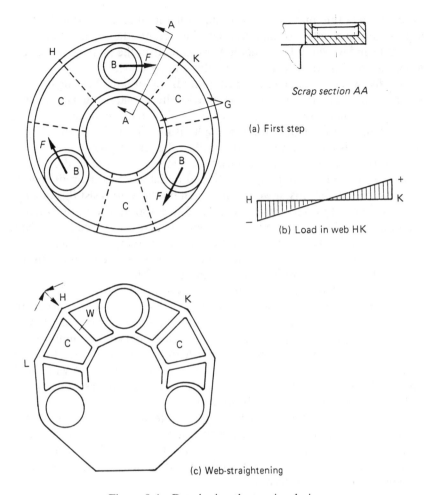

Figure 8.4 Developing the carrier design

clockwise, then the portion of the rib from H to the bearing housing B will be in bending one way, into the paper, such that the rib is in compression, and that from B to K will be in bending the other way, out of the paper, such that the rib is in tension. The load in the rib will be a maximum in tension at K, graduating to a maximum in compression at H, and near the bearing housing at B it will be small. Now the curved rib in compression will tend to bulge outwards, and so be less stiff than a straight one, while the curved rib in tension will tend to straighten, with the same effect.

The magnitude of this effect depends on several parameters. A rib which is thin and tall will be affected more than one which is thick and short. If weight is of no consequence, a thick rib may be the answer. If space is at a premium, a tall rib may be unacceptable. On the other hand, if weight is at a premium and space occupied is less important, then a tall thin rib is desirable and we should consider how we can offset the loss of stiffness due to curvature. Figure 8.4c shows how this can be done.

The disc will have to be reinforced round B, to form the bearing housing. From B, we run ribs straight to H and K: the curvature at B is of little consequence because of the absence of load in the region and the local reinforcement provided by the bearing housing. From H to L the rib runs straight, and is continuous with the wall of the column beneath it.

A problem arises at H and K however. On both sides of H the rib is in compression, so that the point H is forced radially outwards, reducing the stiffness: to prevent this, a web W has been provided to resist this outward load and carry it into the disc. The point J will also need reinforcement in the same way, and the same web can continue across to brace J also. Note however, the loads are not opposed, but in the same sense: we are not providing a short force-path closure in this case. At K the rib is in tension, so that the web W there is providing an outward force.

Wherever a flange in tension or compression turns through an angle, a member needs to be provided to support the resultant load. Consider the box beam in Figure 8.5a, which has a bend in the upper flange at H. With the loading shown the upper flange is in tension, and so for equilibrium at H the webs must exert an upward force R to balance the flange forces F_1 and F_2 (see Figure 8.5b). In a thin-walled box beam, the web will buckle under this load in the fashion shown at (c). To prevent this, a transverse web should be inserted in the beam at H.

The form arrived at in Figure 8.4c is elaborate and expensive to make. It would only be adopted in a case where it was worth spending a lot of money to save a little weight, as in an aircraft gas turbine. Otherwise a simple form like that in Figure 8.4a, with curved flanges of rather greater thickness, would be preferred.

While structural forms like Figure 8.4c are too elaborate for most metal components, they are common in polymers, partly because it is relatively

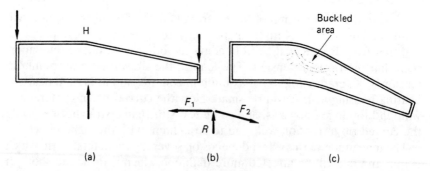

Figure 8.5 Beam with bend in the flange

cheap to mould elaborate shapes in them and partly because lack of stiffness is one of the limitations of polymers. Complicated reticulations of thin flat webs are frequently adopted with these materials for just the sort of reasons which have been discussed in the case of the planet carrier.

8.4 JOINT LOCATION AND DESIGN

Frequently the carrier will be made in one piece, but sometimes the needs of assembly require that it be divided in two somewhere between the ends. From what has already been said about the contraflexure in the columns, it will be best to locate the joint plane at the points of zero bending moment, X, which will be close to the thin disc. The functions required of the joint will be accurate register between the two parts and strength to resist the shear force on the section, which is not very large, and one other role, which is to resist twisting. However stiff the carrier there will be some 'wind-up' or twist, and if there were no resistance to twist in the joint one part would rotate slightly on the other. All that is necessary, however, is to provide sufficient clamping force and friction will do the rest.

8.5 'ALEXANDRIAN' SOLUTIONS

Stiff carriers are heavy and expensive, so it is natural to ask whether there is not some better way. A good start is to apply the principle of least constraint. What we are trying to do with the stiff carrier is to constrain the planet axes to lie parallel with the axis of the gearing as a whole. Because the carrier cannot be infinitely stiff, it cannot be completely successful in this function. But suppose we did not even try to constrain the pinion, but left it free to rotate about the radial axis OA in Figure 8.6a, what would

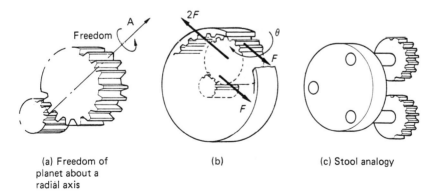

(a) Freedom of planet about a radial axis

(b)

(c) Stool analogy

Figure 8.6 Adding a freedom to the epicyclic

happen? The answer is, it would be constrained by the teeth on the sun and planet to run parallel with them, just as a book laid on the table will be constrained by gravity to lie flat on it. Consider Figure 8.6b, which showed the forces on the planet, F at each mesh and $2F$ in the centre, from whatever structure is going to replace the carrier. The tooth forces are spread uniformly along the length of the teeth, so the resultants F are at midlength, and the $2F$ is also in the centre of the length. All the forces lie in one plane, so they are in equilibrium. Now if the planet were turned through a tiny angle θ in the direction shown in the figure, the planet teeth would bear harder at the left, and so the forces F would move towards that end (remember, the planet teeth are being pushed from behind in the figure). The forces F would then have a moment about the centre tending to turn the planet back, against the rotation θ. Thus the system is stable, and the planet will align itself with the sun and annulus, provided it is pivoted at its centre so the force $2F$ is at the mid-length. The writer applied this concept to the design of a reduction gear for a helicopter gas turbine.

Application of the self-aligning concept

A naive approach to achieving this central support of the pinion is shown in Figure 8.6c. Here the planet carrier is replaced by a 'three-legged stool' with the ends of the legs reaching halfway down the axes of the planets, which are pivoted to the ends of the legs. The 'stool' can be relatively light and flexible, provided the planets are free to turn about the radial axis.

How are the pivots to be contrived? First of all, we could mount the pinion as the outer part of a self-aligning bearing, as in Figure 8.7a. Alternatively, we could use two concentric bearings (Figure 8.7b), the outer one having its inner race stationary and the inner one functioning only to provide self-alignment, a case of separation of function. Yet another way was adopted by the writer in the Napier (later Rolls-Royce)

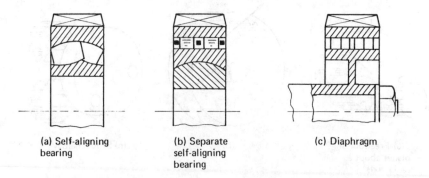

(a) Self-aligning bearing

(b) Separate self-aligning bearing

(c) Diaphragm

Figure 8.7 Options for pivoting the planet

'Gazelle' gas turbine, and is shown in Figure 8.7c, where the inner bearing of Figure 8.7b is replaced by a flexible diaphragm.

Now this idea of pivoting the planets, which brought important savings in weight, was a direct result of applying the principle of least constraint in a place where it was clearly applicable, and, by hindsight, the most obvious thing in the world. Some talk of lateral thinking, but in practice our difficulty is more often that we cannot see what is under our noses. One way of avoiding this nose blind-spot effect is to apply principles consistently.

In this case, having hit on the idea, confidence comes from the crystal clarity of function it displays. If, as in the gas turbine, the carrier is the output, it is being dragged round by three planets, each exerting a force of $2F$, as in Figure 8.6b. The planets in turn are dragged by two tooth forces F which we wish to be evenly spread along the teeth. What is more natural than to secure the planet to the carrier by a central pivot, just as we would secure a horse to a hoe by traces to a swingle-tree fixed in turn by a central pivot to the hoe? We would pivot the hoe direct to the horse were it practicable, just as we pivot the trailer of an articulated vehicle to the tractor.

Consider again Figure 8.6c. Imagine the three planets moving in their common orbit. We wish to take a torque off them, so we plug in the stool-like structure, engaging it ideally by a ball joint in the centre of each planet. Visualise the action, perhaps using the thumb and two fingers of a hand to represent the legs of the stool and imagining them being dragged round. You should now have that insight of which Torroja spoke in regard to structures, of which he said that the designer should grasp the working with the same clarity as that with which he understood the forces that caused a stone to fall to the ground or drove the arrow from a bow.

8.6 THE CHOICE OF EMBODIMENT

Having secured the concept, turn to the embodiment. Why choose the diaphragm rather than one of the self-aligning bearings? The self-aligning bearing of Figure 8.7a would work at low speeds and moderate loads, but is not suitable for the high speed and strong centrifugal field in the gas turbine application. Indeed, the centrifugal field later gave trouble with the roller bearings, causing the rollers to fly across the pockets in the cages when they came to the unloaded part of their orbit and were released. The resulting blows eventually fatigued the standard cages, so that they had to be redesigned in a stronger form. The objections to 8.7b, the separate self-aligning bearing, were based on the fact that there is no relative rotation and a steady load, so there could be no squeeze or wedge action to generate a hydrodynamic film.

To provide a hydrostatic bearing would have been possible, but there was no available source of oil at a high enough pressure. In the absence of a film, there could have been a high residual torque on the bearing without it realigning, permitting a considerable variation in tooth loading along the length.

Biasing the diaphragm

The diaphragm requires no lubrication and cannot stick. On the other hand, it has to be flexible enough and yet strong enough to carry the very high loads involved. A first thought would be to make the diaphragm very flexible, much more so than the carrier, now reduced in structural terms to a 'three-legged stool', as in Figure 8.6c. However, this is difficult to achieve and involves an arbitrary decision. This is a case for a little more insight, starting with a reference back to what we seek to achieve.

Figure 8.8 shows a radial view of one planet inner race and diaphragm and the part of the 'stool leg', L, which supports it. Under the load $2F$ from the planet, L turns anticlockwise through θ. If the diaphragm is put h to the right of centre of the planet, then when the tooth load is uniform, there will be a moment $2Fh$ turning the planet clockwise. This clockwise moment $2Fh$ will bend the diaphragm, resulting in a clockwise rotation on top of the anticlockwise rotation θ. By giving h the right value, the two effects can be contrived to be equal and opposite.

The most insightful way of describing this is to say that we balance the flexibility of the 'stool' or spider and the dimension h such that a load at the centre of the planet produces *deflection* but no *rotation*. We do this by taking the load path to the right first in Figure 8.8, before bringing it back

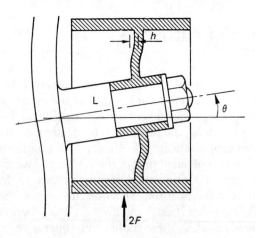

Figure 8.8   Locating the diaphragm

to the left, the flexibilities on right and left balancing one another. Consider the structures shown in Figure 8.9, where the parts shown by heavy lines are rigid.

Figure 8.9a represents the 'three-legged stool' design without the diaphragm, idealised as a beam. The part AB, a cantilever, represents the flexibility of both the 'legs' and the 'seat' of the stool, so that A must actually be well to the left of the 'seat' or disc to give representative flexibility, since most of the bending is in the 'seat'. The deflection causes slope in BC, and hence in the planet, as shown by the broken lines.

Figure 8.9b represents the design with the diaphragm, with AB extended to B' and the diaphragm, B'C, made flexible. By the right choice of BB', which is h in Figure 8.8, the slope at C can be made zero.

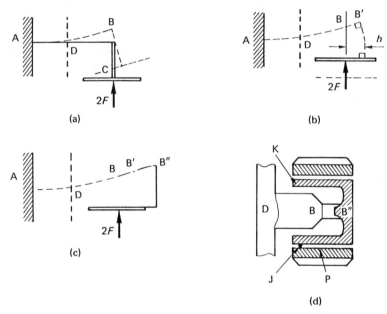

Figure 8.9 Self-aligning planets; flexible pins

8.7 ANOTHER SOLUTION

Figure 8.9c shows another possibility, based on the fact that in Figure 8.9b, the length BB' of the 'leg' is in reverse bending, which contributes slightly to removing the slope of the planet. By extending AB even further, to B", the flexible diaphragm can be dispensed with altogether. This ingenious solution is due to R Hicks.

Instead of the flexible diaphragm, the 'leg of the stool' is just extended until a central load on the planet causes no slope, only deflection. In practice, a little more is needed. The structure resembles that of a spring in the bathroom scales (Chapter 2), and if AB″ were a uniform beam, B, the point opposite the load 2F, must be the midpoint of AB″. This would give much too long a leg, since the effective disc is at D (see Figure 8.9c). However, the leg need not be uniform: the part BB″ where the correcting bending is taking place can be made more flexible than the part DB (remember, the part AD is imaginary and simply represents the flexibility of the disc). As with the bathroom scales, the bending moment in BB″ is zero at B and a maximum at B″, but the proportions of the parts in the epicyclic gear leave little scope for using taper to match the section modulus as was done in the case of the scales.

Figure 8.9d is a rough sketch of an embodiment of the beam self-aligning arrangement of Figure 8.9c: it takes no account of the needs of manufacture. Here the planet P runs on a hydrodynamic bearing J. Note the change of section at B, which makes the length BB″ relatively flexible by just the right amount to correct the slope at B due to bending in the disc and the length DB.

The inner part K of this bearing, the disc D and the 'beam' or pin DBB″ cannot practically be made in one piece, so a joint must be made. The idea developed before, that a joint in a member in bending such as DB″ should be placed at a point of contraflexure, such as B, where there is no bending moment, can be applied, but this would not be very helpful from the manufacturing point of view.

Figure 8.10 shows a possible joint at B, following the philosophy of the bevel gear mounting in Figure 5.8. Notice that the register at B can be either a flat face and a spigot, as shown in Figure 8.10 in the top half, or a cone as in the bottom half (shown in a circle). The whole joint is secured by

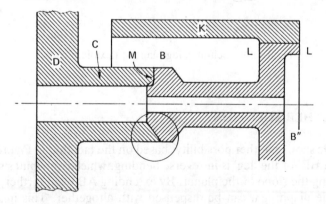

Figure 8.10 Joints in pin

a single large bolt from B″ to D. This special bolt is not shown, but it should have a spigot location at D to centre the head.

With the arrangement of Figure 8.10 it is convenient to make the parts BB″ and K separately, fixing them together permanently at LL, by brazing or welding, say. To ensure sufficient accuracy, it would be necessary to finish machine the surfaces at B and K afterwards, preferably all at the same setting. Unfortunately the surfaces at B are difficult to come at because of K, which might be a reason to move the joint to C (Figure 8.10) in spite of the large bending moment.

The central bolt has to be able to prestress the joint face adequately to carry the bending moment. It would help to make the tubular portion at C as large as possible, with a narrow abutment, for the reasons discussed in Chapter 3 in connection with the big end and the stator blade fixing. A very simple approach would be to put such a joint at C and another at B″, instead of the permanent joint at LL.

In deciding between alternatives of these kinds, calculations of stress, deflections and costs are needed. Decisions will depend on the particular proportions of the gears and the size of the loads they carry, and to a certain extent on the resources available in the company concerned. In a general way, space considerations mean that the diaphragm solution is only useful for large ratio epicyclics where the planet diameter is large compared with that of the sun. With small ratios, giving large suns and small planets, there is no room for a diaphragm but there may be room for a pin, particularly if the face width is large.

8.8 OTHER ASPECTS OF EPICYCLIC GEARING

There are other interesting problems associated with epicyclic gearing besides the planet carrier. The alignment of the teeth is also upset by the flexibility of the sun pinion, which twists or 'winds up' under the torque in it, causing heavy loading on the teeth at the input end. Figure 8.11a shows the effect of twist, which will increase parabolically from the free end A, where the torque, and therefore the twist per unit length, is zero, to the input end B, where the torque, and therefore the twist per unit length, is a maximum. Twist per unit length is the gradient of twist and, being proportional to the torque, would increase linearly along the tooth if the tooth load were uniform: since the twist tends to concentrate load towards B, the effects are even worse. The exact solution is given by a second order differential equation, comes out in hyperbolic functions, and involves the stiffness of the gear mesh.

One way of improving this situation is to apply some bias, by making the pinion with a slight helical angle (if the gearing is spur gearing) or a slightly different axial pitch (if the gearing is helical). In the case of spur gearing,

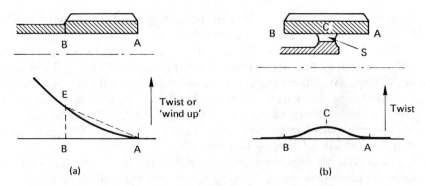

Figure 8.11   Twist in sun pinion

this can be done by applying half the design torque, the same way, to the pinion while it is being ground to its finished form. Then the deviation from straight when running at the design torque will be the same as in Figure 8.11a, but based on EA as axis. This way will only work well at one torque, of course.

Another way is illustrated in Figure 8.11b. The drive is fed to the pinion half-way along, at S, by means of an inner shaft (quill-shaft) connected to the pinion by splines or serrations at S. This way, the twist builds up from both the ends A and B of the pinion towards the centre C, as in the figure, reaching only about one-quarter the value.

In Section 4.4 it was shown that with three planets the loads could be balanced between them by allowing a member to float transversely. In addition, and especially with more than three planets, the annulus gear deflects radially under the tooth loads and helps to balance the loads, working rather like an elastic band round a bundle of pencils (only in the annulus, the springiness is in bending, not in stretching).

Summary

This study of the epicyclic gear has covered only a few points, but they are perhaps the ones most relevant to design in general. In particular, they relate to the great group of problems designers face in using materials of limited stiffness, problems which are met in many other fields. For instance, the design of rolling mills for rolling sheet steel involves difficulties of a similar kind, because the bending of the rollers tends to make them roll sheet which is thick in the middle, and the ingenious shifts adopted to overcome this problem are all in the spirit of this chapter.

9

Hydraulic Pumps

9.1 THE SWASH-PLATE PUMP

There can be few areas of engineering design where more ingenuity has been displayed than in that of hydraulic pumps. While internal combustion engines, for example, have tended to converge on a few preferred types, virtually all based on the crankshaft and piston-in-cylinder principle, high-pressure displacement pumps have proliferated forms, many of which have successfully occupied distinctive niches in the market. The field is rich in examples for the engineering designer, both of problems and of methods of solution.

Consider the swash-plate pump shown in Figure 9.1. The cylinder block rotates, carrying bodily with it the pistons, the ends of which are connected by ball joints to slippers running on the swash-plate itself. This does not rotate, so that the slippers and hence the pistons ride up and down on it as they orbit round the pump axis. Porting in the valve plate is arranged so that each cylinder is connected to the delivery port on the upstroke of the piston in it, and to the inlet port on the downstroke.

By reducing the angle of tilt θ of the swash-plate, the stroke and hence the swept volume of the pump may be reduced: to this end, the swash-plate is pivoted about an axis through O normal to the paper and provided with some control arrangement, often a small hydraulic cylinder acting against a spring, as shown.

On the up-stroke, the slipper is held against the swash-plate by the pressure in the cylinder: on the downstroke it will be left behind unless some *closure* provision is made. The usual means is a plate with holes in it (P in Figure 9.2): the slippers move relative to the holes, which are large enough to allow the movement, but have projecting ledges which always

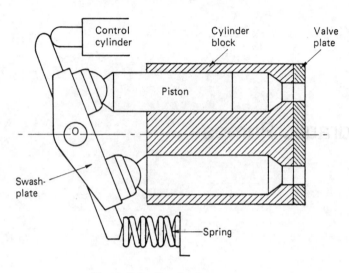

Figure 9.1 Swash-plate pump

remain trapped under the plate, so providing a *form closure*. Once springs were often used to provide *force closure*, but these were scarcely strong enough, liable to break, occupied valuable space and increased losses.

9.2 VALVE PLATE DESIGN

An excellent example of clarity of function and kinematic design is provided by the valve plate. This is basically a simple disc containing two banana-shaped ports (see Figure 9.2), which rests on the end of the cylinder block but does not rotate with it. The ports come opposite the holes in the tops of the cylinders. One is opposite the 'upstroke' side of the swash-plate where the pistons are rising, and so this side is the delivery or high-pressure side. At the top of the stroke the hole in the top of a cylinder passes the dead space between one 'banana' and the next, and then, as the piston starts the downstroke, comes opposite the inlet or low-pressure port. Simple porting arrangements of this kind are often possible with pumps.

Consider the degrees of freedom of the valve plate in the casing. The principle of least constraint says that these should be as many as possible. We have just seen that it needs to be prevented from rotating about the axis of the pump, and it is clear also that it needs to be held concentric with the cylinder block, that is, its two translational degrees of freedom at right angles to the axis of the pump must be suppressed. For the moment, let us assume that it is not restrained in the other degrees of freedom, those of

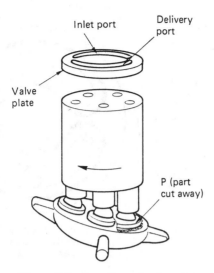

Figure 9.2 Function of valve plate

axial motion and rotation about axes normal to the pump axis, which may be called the freedom to float and the freedom to tilt.

Now consider the sealing problem. Oil is going to leak through the gap between the valve plate and the cylinder block from the tops of the cylinders which are pumping at any instant. In effect, the contact between the rotating cylinder block and the stationary valve-plate is a hydrostatic bearing, fed by the pumping cylinders. The whole banana-shaped high-pressure port is full of oil at delivery pressure which forces its way into the clearance between valve plate and cylinder block, dropping in pressure in the narrow space as it does so (see Figure 9.3a, which shows a typical constant pressure line).

In order to keep the leakage low, the valve plate must be pushed down on to the cylinder block, and it must be free axially to do so. It must also be free to tilt so it can conform angularly with the face of the block, confirming the tentative decisions made on the basis of the principle of least constraint. The degrees of freedom are the same, and for the same reasons, as those required by the tap washer in Chapter 1 (except the tap washer is also free to rotate about its axis).

Finally, it is necessary to seal the gap between the valve plate and the casing. Because of the need to allow for float and tilt, this gap must be relatively large. Fortunately, there is no relative rotation, so that O-rings in grooves can be used. The O-rings need to enclose the ports, and their dimensions must be chosen so as to force the valve plate towards the cylinder block with just the right resultant force in just the right place: the centres of pressure on the two sides of the valve plate need to be as nearly

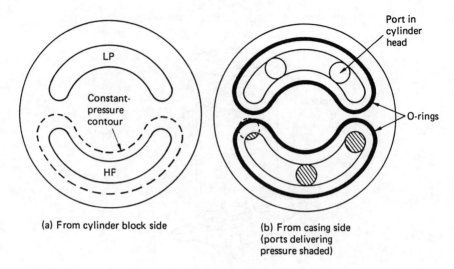

(a) From cylinder block side

(b) From casing side (ports delivering pressure shaded)

Figure 9.3 Pressures on valve plate

opposite one another as can be contrived. Roughly, this means that the O-rings should be in banana-shaped grooves lying roughly half-way between the edges of the ports and the edges of the plate. Notice that an O-ring is needed round the inlet or low-pressure port also because this will generally be at above atmospheric pressure.

Completing the design

There is a lot of work in completing the design. The leakage flow and hence the pressure distribution in the clearance between the valve plate and the cylinder block, when that space is parallel, will be governed by the Laplace equation. The O-rings must be kept within their working range with the worst combination of tolerances, and allowance must be made for differential thermal expansion of parts between whatever location (thrust) bearing is provided for the cylinder block and the face of block.

Notice that the cylinder block is subject to two large axial forces, an upward force from the cylinder heads of the pumping cylinders (less the delivery holes) and a downward force from the space within the banana-shaped O-ring round the delivery ports (less the delivery holes). These forces roughly balance, and it is desirable to decide which way the load should be. This can be controlled, for instance, by choosing the effective diameter of the valve plate on the cylinder block side by relieving the surface, so that the leakage pressure will drop to that in the casing at the beginning of the relieved part.

HYDRAULIC PUMPS 179

The design of the valve plate is a good example of clarity of function. The situation is a complicated one, and the designer decides how the parts should work and sets out to ensure they will do so – in this case, by the choice of dimensions and, eventually, the control of tolerances. He determines that the valve plate should be forced against the cylinder block, and dimensions the parts so that it is. He determines which way to load the location bearing, and arranges the areas accordingly, and so on.

9.3 ELIMINATING THE VALVE PLATE

It would be possible to design a pump in which the banana-shaped ports were in the casing itself, and it might be thought that this would be cheaper. But it would be difficult to ensure parallelism with the top of the cylinder-block and to machine a smooth enough face at the bottom of the hole in the casing, and if a split plane were located at the level of the ported face, the alignment difficulties would multiply. Also, it is desirable to use for the port face a bronze or similar material which will form a good bearing pair with the steel or cast iron of the cylinder block, but to make the whole casing of bronze would be very expensive. On the other hand, to make the relatively small and compact valve plate of bronze is not too expensive. But above all, to make the ports in the casing would make it difficult to keep the efficiency of the pump high. Hydraulic pumps commonly have efficiencies well into the nineties per cent.

The separate valve plate is an example of 'separation of function' (Chapter 1); the tasks of 'contain pressure' and 'provide valve ports' are divided between two components, giving the advantages which have been discussed. Notice that the valve plate is simply dropped into its recess in the casing, taking care to engage the pin or other feature provided to restrain it from rotating. This 'drop together' construction is rather characteristic of kinematic design, where parts only touch on certain surfaces, carefully chosen and often quite small. The grooves for the O-rings can be in casing or valve plate, but it is cheaper to machine them in the valve plate, especially since they will be profile milled like the ports themselves, and all the work can be done in one setting.

9.4 VIRTUES AND LIMITATIONS OF THE SWASH-PLATE PUMP

Before leaving the swash-plate pump to look at another kind, let us look at its virtues and vices. It is capable of high efficiency, it is variable in stroke quite readily, it is fairly compact and light for its power. It is an excellent design, but it has, like all designs, its faults and limitations. Chief among these is the side-thrust on the slippers (see Figure 9.4).

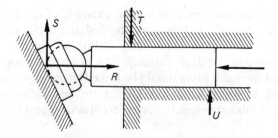

Figure 9.4   Side-thrust in swash-plate pump

The net thrust on a slipper is normal to the swash-plate, and can be resolved into a component R along the cylinder (see figure) and a component S at right-angles to the pump axis. The side forces S on the working pistons, multiplied by the pitch radius of the cylinders, constitute the driving torque of the pump: they are balanced by forces T and U (see figure) on the pistons, that is, the driving torque is transmitted from the cylinder block to the swash-plate where it is reacted by the bending moment in the pistons, an inelegant method which should displease the critical designer. At a given swash angle θ and delivery pressure p, the force S will be a maximum at midstroke but the force T will be greater when the slipper is further away from the block, a little earlier in the stroke. It is hard to ensure good lubrication under the load T when it is high, and a breakdown would result in failure.

High swash angles increase S and also increase the ratio of T to S, and thus there is a great incentive to keep the swash angle small (typically, not more than 20°). This limits the capacity of the pump and so the ratio of swept volume to overall volume which we saw as a criterion of excellence in the quarter-turn actuator. The low swash angle also means slightly lower efficiency, since with a bigger swash angle the leakage losses would be no greater, but the useful power would be increased.

There is another form of hydraulic pump which resembles the swash-plate pump but does not have the side force problem, the bent-axis pump.

9.5   THE BENT-AXIS PUMP

The bent-axis pump shares several features with the swash-plate pump, notably the rotating cylinder block and the porting arrangement which are virtually the same. However, the bent-axis pump has connecting rods to its pistons, which are driven by an inclined drive shaft as shown in Figure 9.5. Because of the bend in the axis (the angle θ in the figure) the cylinder block cannot be driven directly by the drive shaft, and so some other means must be adopted: in Figure 9.8, bevel gears are shown, but other methods also can be used. In any case, the drive shaft and the cylinder block rotate at the

HYDRAULIC PUMPS   181

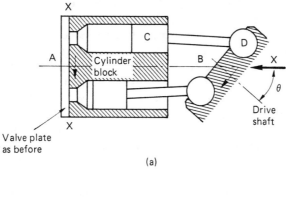

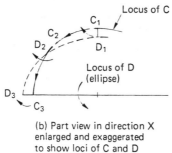

(b) Part view in direction X enlarged and exaggerated to show loci of C and D

Figure 9.5 Bent-axis pump rotating elements and valve plate (drive to cylinder block omitted)

same speed about their respective axes, the pistons move up and down in the cylinders, and oil is pumped through the delivery port just as in the swash-plate pump.

It is always desirable to have a clear understanding of the 'force paths' in any machine. In a pump, there is a torque in the drive shaft and this must be balanced by a reaction torque from the casing. In the case of the swash-plate pump, it came from the swash-plate itself. In the case of the bent-axis pumps, it comes from the pressure from the delivery port on the casing. This load is off-centre, so it can be regarded as a central force combined with a moment, M say, about the axis XX in Figure 9.5a. There is an exactly similar moment in the swash-plate pump, but that does not have a component about the axis of the drive shaft: in the bent-axis pump M does, because of the bend; it has a component $M\sin\theta$ which balances the input torque.

In the figure, the pumping cylinders are those towards the front: as they move down, the heavy compressive forces in the connecting rods have components tangential to the driving shaft, resisting its rotation. However, there is here no undesirable effect to correspond to the side load T in the swash-plate pump, and the angle θ, corresponding to the swash angle, can be increased to about 40°, giving a stroke twice as long and a corresponding improvement in the swept volume/overall volume ratio.

In the nature of things, we expect to find some disadvantages in the bent-axis pump. One is that it is awkward to vary the stroke. Since mechanical drives are hard to bend, it is easiest to vary θ by pivoting the cylinder block end about the axis through O normal to the paper in Figure 9.5a. That means that the oil inlet and outlet on the left are moving in space, and either flexible hydraulic lines must be used or passages must be made in the cylinder block housing to lead the oil in and out via hollow pivots on the 'bending' axis, at O. In consequence, the bent-axis pump makes but an awkward variable-stroke machine and its use is generally confined to fixed-stroke versions.

It is also more complicated than the swash-plate pump, with its pistons ball-jointed to connecting rods and the need to drive the cylinder block from a drive shaft at an angle. In Figure 9.5a a bevel drive is shown, and this is pure and above reproach, but somewhat expensive and likely to increase the bulk. Usually some form of constant-velocity joint is used, tucked in between the connecting rods and the cylinders, where it is difficult to accommodate.

Kinematics

The kinematics of the bent-axis pump are interesting. The ball joints between the connecting rods and the drive shaft vary in distance from the axis AB of the cylinder block (see Figure 9.5a), being closest at the ends of a stroke and furthest away at the middle of a stroke. If the pitch circle radius of the cylinders is made equal to the average of these extreme distances, the angle made by any connecting rod to the axis AB remains constant, at about 3°, while the plane containing the connecting rod and the axis of the cylinder rotates at the same speed as the pump but in the opposite direction. This may be appreciated from Figure 9.5b, which shows the path of the ends C and D of a connecting rod, viewed along the pump axis AB. The path of the piston end, C, appears as a circle and that of the other end, D, as an ellipse. Three successive positions are shown: the projections D_1C_1 etc. seen in this view remain constant in length, while rotating the opposite way to the cylinder block. This result is easily shown, for example, by appeal to a well-known construction for an ellipse.

9.6 AN ELEGANT BENT-AXIS PUMP

A particularly interesting design of bent-axis pump, by Volvo, illustrates several general aspects of design. Consider first the closure aspect: unless the low-pressure or inlet side of the pump is always to be pressurised, then it will be necessary to provide some closure to the ball joints at the ends of the connecting rod so that the piston will be drawn down the cylinder on the out-stroke and not left at the top. This can be achieved in a form closure by ensuring that the socket in which the ball fits embraces it over more than a hemisphere.

This can be done as shown in Figure 9.6a, by making half the socket in the drive shaft and providing a cover corresponding to the northern tropics, as it were, of the socket. All these covers can be made in one piece, but that piece still has to be made and fastened to the drive shaft, and there is an assembly problem, since the ball D will not go through the hole in the cover. However, only a part of the cover is required, so that a slot can be made, at E say, to pass the connecting rod stem. This 'keyhole' solution is useful in many situations, such as on some poppet-valve stems (see Figure 9.6b), where the keyhole plate is slightly dished so that the spring prevents it moving off-centre and disengaging itself. Unfortunately, the keyhole form gives a large but weak structure which is often unacceptable.

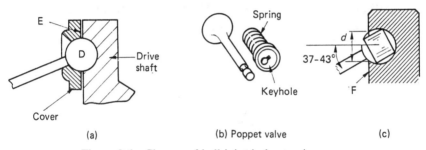

Figure 9.6 Closure of ball joint in bent-axis pump

The elegant solution used by Volvo takes advantage of the fact that in a bent-axis pump where θ is fixed, the connecting rod *always makes a large angle with the axis of the drive shaft*. In their case, θ is 40°, and as we have just seen, the angle of the connecting rod to the drive shaft will swing $\pm 3°$ or so from this value. It does not matter if there is no closure when the connecting rod is parallel to the drive shaft. Figure 9.6c shows the Volvo solution, where the 'tropical zone' of the ball has been machined down to a cylinder, of diameter d, and the socket embraces the 'southern hemisphere' and the 'northern tropics', up to the latitude where the diameter is d. Thus the ball can be assembled into the socket with the connecting rod axis parallel to the drive shaft, but is locked once the axis is bent.

The area of the locking surface is small, but then only a small load is required to withdraw the piston. The heavy loading occurs during the pumping stroke, when the whole of the 'southern temperate and polar' regions are available to share the load: this is a squeeze film application (Chapter 4) and it is important to ensure the oil is there to squeeze.

This elegant and simple solution of providing form closure for the ball joint is made possible by the fact that it is always heavily bent in service, in others words, there is something unusual about the specification which is less demanding than usual: the joint never has to work in the straight position, or anywhere near it. It is very easy to overlook such 'easements' or relaxations, and so to fail to take advantage of them in design. The designer is usually only too ready to recognise extra difficulties in a new problem, but not nearly so quick to take advantage of respects in which it is easier than usual.

The piston design

Another feature of this pump is the pistons. These are spherical, and integral with the connecting rod and the ball of the ball joint (see Figure 9.7). Ideally, the spherical piston would touch along a circumference of the cylinder as shown by the broken line, but in practice an equatorial piston ring is fitted to improve the sealing. The top of the sphere is cut away so that the length of the cylinder can be reduced a little.

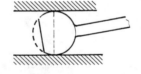

Figure 9.7 Spherical piston

The casing

There is another good basic lesson to be learnt from this pump, which relates to the casing, shown in Figure 9.8. It is made in two pieces bolted together, and there is no other joint in it. A bolted joint can be made to serve other purposes than assembly, for instance, clamping a bearing outer race in position. However, in this case it also needs to provide a seal to prevent oil leaks from the pump. Now the clamping action of the joint cannot be used both to compress a seal and to clamp a bearing. However, a seal can be made on a diameter as well as on a face, and in this case O-rings in circumferential grooves are used for sealing, leaving the joint free to tighten on the bearing race A and the spacer B beside it.

This is a very elegant and economical arrangement. The bearings A and C are of the taper roller type, preloaded one against the other by the shaft

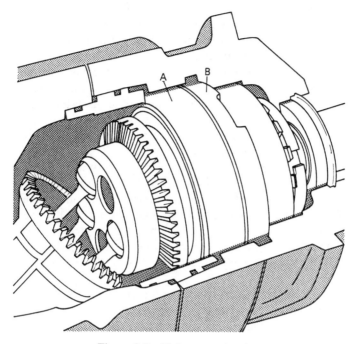

Figure 9.8 Volvo pump casing

and the nut on the end of it acting like a large bolt and nut. When this has been done, the shaft and bearing assembly can readily be mounted firmly in the casing, using the single bolted joint to clamp one race in place. The whole is then secure, on the same principle as was shown in Figure 3.2a. Notice that only A and B are clamped axially, and only A fits closely in the bore of the casing, that is, locates in translation.

9.7 COMBINING THE VIRTUES OF SWASH-PLATE AND BENT-AXIS

The two kinds of pump we have studied have different virtues. The swash-plate pump is well-suited to a variable stroke version, but the bent-axis type has a larger swept volume for its outside dimensions, because the angle of bend can be greater than the swash angle. If the problem of the side thrust on the slipper could be overcome, the swash-plate pump could have double the maximum stroke, and be smaller and more efficient.

The bent-axis pump takes the side thrust in the drive shaft. If the swash-plate pump could do the same, it could use a large swash angle. A first step would be to replace the separate slippers by a solid ring, corresponding to the solid disc of sockets on the drive shaft of the bent-axis pump. This arrangement is shown in Figure 9.9. All the side thrust will now be taken by this ring R, so that instead of support for several pistons, just one part must be provided for. Because of the kinematics of the arrangement, with the sockets evenly spaced on a circle in a plane inclined to the axis of the cylinder block, connecting rods with joints at the ends would be needed, just as in the bent-axis pump, but this complexity is acceptable. The problem remains, of how to connect this ring, which would drive all the pistons through piston rods, to the drive shaft.

Consider the degrees of freedom of the ring, which will be six before any attachments are made to it. It will be flat on the swash-plate, which reduces it to three degrees of freedom – two of translation parallel to the plane of the swash-plate and one of rotation about an axis normal to the plane, just the usual three for a body moving in a plane. We need to remove the two of translation and constrain the rotation to be with the shaft. Should these three constraints be provided through the shaft or the swash-plate, or even the casing?

At this point some readers, recalling the epicyclic gear of the previous chapter where letting everything go free seemed the right thing to do, may ask whether all these constraints are necessary. In Figure 9.9, the cylinders A and B are pumping, so there are large compressive forces in the two connecting rods marked with arrows. If unrestrained, the ring R would slide off in the direction y (see figure), so clearly a constraint in y is needed. The case of the freedom x is more difficult, but not much: the ring is unstable in this direction under the thrusts of the working connecting rods, because a slight movement x produces a slope in the rods which causes the

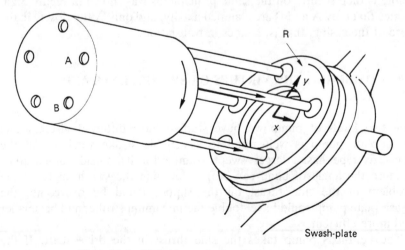

Figure 9.9   Hybrid swash-plate/bent-axis pump

thrusts to have a component in the same direction, and so without a constraint in x the ring would shoot off to one side or the other. The effect is like that of stepping on a piece of soap, and is all part of the essential instability of compression members (Chapter 2).

The rotational constraint, because it actually drives the ring R and hence the pump, must come from the shaft. The easiest way to provide the constraints on x and y would be a simple journal bearing between R and the swash-plate, a pivot, in effect. The connection to the shaft should then be a pure torsional constraint, following the Principle of Least Constraint. Is there anything in the design repertoire, the stock of known solutions, which fits the bill?

Figure 9.10 shows a plane mechanism which imposes a pure torsional constraint: the member B can translate in the plane relative to member A, but cannot rotate relative to it. A 3-D pure torsional constraint is likely to be even more elaborate and awkward. For instance, we could add a universal joint to B to attach a third component C, and then a slide to connect a fourth component D, and we would have a pure torsional constraint between A and D, but at what a price in complexity!

Consider the alternative of constraining the ring relative to the central drive shaft in rotation *and* in x and y. Here there is a suitable component in the repertoire, the 'plunging joint' mentioned in the discussion of car suspensions in Section 5.9. This is a joint used to permit bending in a shaft, like a universal or Hooke's joint. Unlike a universal joint, however, it transmits rotation at constant velocity (it is a constant velocity joint) and it also permits axial sliding (hence the 'plunging', which sets it off from other constant velocity joints).

Unfortunately, hydraulic machines handle very large torques in very small volumes. In all torque-producing devices, as was seen in the quarter-turn actuator in Chapter 7 (equation (7.3)), the torque per unit volume is proportional to the pressure, and the pressure may well be even a hundred times higher in hydraulic machines than in pneumatic ones. 'Plunging' and similar joints transmit torque via balls in grooves, and

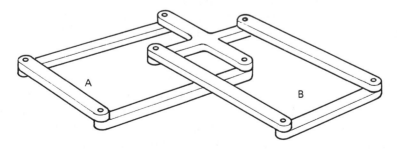

Figure 9.10 A plane torsional constraint

as was seen in Section 2.7, the contact area in such cases is very small, and so even though the allowable stresses are very high, it is not possible to squeeze enough torque capacity into the restricted space available.

Generally speaking, joints which provide more than one degree of freedom will rely on weak line or point contacts. To obtain high torque densities it is usually better to nest joints giving one degree of freedom each. To give an example, the crank in the quarter-turn actuator in Figure 7.1d is connected to the pin P on the piston rod by a joint giving two degrees of relative freedom, sliding along the slot and turning about the pin. This could be achieved by a simple round pin in a slot, as in Figure 9.11a, but the contact between parts would be along a 'line'. By fitting a block between the slot and the pin, as in Figure 9.11b, contact at the two joints in series, the sliding one and the turning one, is over an area in both cases, and the load capacity is increased by more than an order of magnitude.

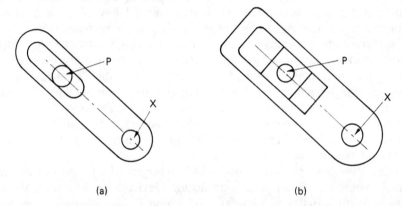

Figure 9.11 A joint with two degrees of freedom

The inventor of the new hydraulic machines based on this idea, Robert Clerk, decided to use three ball-jointed links to connect the ring R to the drive shaft, after the pattern shown in Figure 9.12. He calls this arrangement the tri-link. It follows the principle of using surface-to-surface contact (lower order pairs) to obtain high load capacity, and this involves using six ball joints.

It is instructive to count degrees of freedom. The need was to remove three from the ring R, namely, x, y and rotation relative to the drive shaft. This was done by adding three links, each with its own six degrees of freedom, giving eighteen more. Each of the six ball joints removes three, so that it might appear these three links remove exactly as many degrees of freedom as they add, with no constraint left over. Intuition says this is not so, and intuition is right, because each link is left with a degree of freedom,

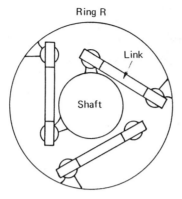

Figure 9.12 Principle of tri-link joint

rotation about the line joining the centres of the ball joints at its ends. Since there are these three additional freedoms now, and since it has just been shown there is no net change, three other degrees of freedom must have been removed; these are those of translation and rotation relative to the drive shaft.

Other joints of high torque capacity might be considered. A universal or Hooke's joint would not be acceptable because it would locate the ring axially, a function already filled by the swash-plate. A Hooke's joint has two degrees of freedom, and so removes four, so that there would be a duplicated constraint, in conflict with the Principle of Least Constraint and likely to lead to conflict in practice. By allowing the Hooke's joint to slide axially on splines or serrations, the desired freedoms would be met, but at high torques such joints tend to lock due to friction. A few large splines with hydrostatic bearings on the mating faces might be considered, but this idea is not appealing. The earliest machines, however, to use this principle of a swash-plate plus connecting rods and a constrained ring R did use a Hooke's joint (Williams–Janney).

Other features of the Clerk Trilink machines

Two other features deserve attention here. The first is that the cylinders have thin liners, as in Figure 9.13, open to the interior at the head end so that the pressure behind them is the same as in the cylinder. During the pumping stroke, the high pressure in the cylinder expands it, and without a liner, this increases the clearance round the piston and hence the leakage. The liner of the Clerk design, however, is subject above the piston to equal pressure inside and out, and so does not expand (it will in fact contract very slightly, as does the piston), and there is no loss of fit. Below the piston, the high external pressure does squash the cylinder inwards, causing it to

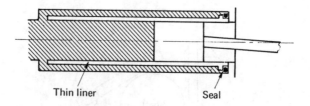

Figure 9.13   Cylinder liner (shaded area pressurised)

contract behind the piston. During the down or suction stroke all the pressures are low and deformations are small.

This is an example of the separation of functions, the two functions being the prevention of leaking past the piston and the prevention of bursting. In other machines both functions are performed by the cylinder, which not only stretches, but stretches irregularly round its circumference, because the wall-thickness is not uniform. In the Trilink machine, the cylinder has only to resist pressure, the role of preventing leakage being taken over by the liner, which is largely unstressed and of uniform thickness.

The other feature is that the design is thoroughly 'kinematic', in details like having self-aligning bearings. Such designs can run well even when the deflections in the parts are relatively large, as was seen in the case of the epicyclic gear in Chapter 8.

General remarks

Hydraulic machines are interesting examples of elegant and sophisticated design. They are remarkably efficient, and they pack a great deal of power into a small space. The three related kinds that have been discussed are among the most advanced, but there are many other fascinating varieties. There is a large family of radial piston machines, some with a single crank pin, others with a cam ring round the outside of outward-pointing cylinders: many show great subtlety and refinement. They are a rewarding study for anyone interested in functional design, full of conflicting considerations among which engineers of great originality have made many inventive and shrewd (and less shrewd) sets of choices.

10

Miscellaneous Examples

10.1 CONNECTING RODS

In Chapter 3 the design of the joints that have to be made in connecting rod big ends was considered. In large marine engines there is an added complication because the crank pins are large and there is a need to withdraw the connecting rod through the cylinder, thus limiting its lateral extent. Figure 10.1 illustrates the problem, and shows different forms of split line, or strictly, split surface, between the two parts of the big end which have been adopted to solve it.

Students set this problem as an exercise tend to see at once that the split plane can be angled to overcome the difficulty. They do not recognise at first that the split surface need not necessarily be a *plane*. Once this is pointed out, they often go on to draw unacceptable split surfaces: having

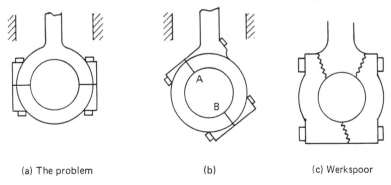

(a) The problem   (b)   (c) Werkspoor

Figure 10.1   Marine diesel big ends

started with too narrow a category, they then switch to too large a one. It is a useful exercise to draw up a set of conditions for a suitable split surface.

First, we assume that it should be a prismatic or cylindrical surface, that is, it can be generated by a straight line moving so as always to be parallel with the axis of the crank pin. This is an arbitrary decision which we should review later.

The whole object of the joint is to enable the big end to be assembled round the crank pin. It follows that the intersections of the split surface with the bore, A and B in Figure 10.2a, must be diametrically opposite one another, since the angle of embrace of any part must be 180° or less to allow it to be drawn off the crank pin.

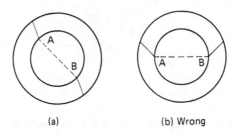

Figure 10.2   Specifying the split

Beyond this, it is desirable that the splits in the big end should be *generally flat* and *parallel* one with the other. This condition is imposed because of the need for proper tightening combined with reasonable tolerances. With an arrangement such as that in Figure 10.2b, the time-honoured principle, 'never make two faces at once', is violated. The positions of the bolts are not shown, but they cannot be both parallel to one another and perpendicular to the faces they prestress. There is a little of 'making two surfaces at once' about the split shown in Figure 10.2a. Tight tolerances are necessary on the offset between the two parallel split planes, but in practice this design is used and has proved acceptable. The bolts can be parallel and normal to the faces. The Werkspoor design of Figure 10.1c has also proved acceptable, even though the bolts are not perpendicular to the faces.

The term 'generally flat' was used because the faces are usually serrated: the serrations only affect the problem in that, in a design like that in Figure 10.2b, they prevent lateral sliding during closure, an extra constraint which adds to the difficulties.

The princess and the pea: the Wasa design

The next example shows very beautifully the kind of structural insight the designer should strive for. It hinges on a very lucid appreciation of the function of the big end, which is to transfer the load from the connecting rod into the crank pin via an oil film. This particular load transfer is critical: it determines the allowable working pressures in the cylinder (the mean effective pressure), and hence the power rating of the engine. We want the load to be spread over as large an area of the crank pin as possible, which means that the clearance between pin and bearing must be suitable over that area. This is one of that large category of *load diffusion* problems, which includes also the problem of diffusing the load of a sleeping princess into a bedstead.

It was pointed out in Section 3.4 that the big end functions as a ring beam. Massively deep and stiff as this beam is, it still deflects under the great loads on it an amount which is significant compared with the thickness of the oil film. How should we approach the problem of designing the connecting rod so as to obtain a good load distribution?

Fortunately, we know precisely how we would like the load to be distributed on the crank pin/big end interface. The heavy load occurs when the piston is near the top of the stroke and it is a compressive force in the connecting rod. We would then like the load in the bearing to be spread through a large arc with a nearly uniform distribution in the middle, falling off towards the ends, as in Figure 10.3. How should we take hold of the big end on the other side in order to achieve this desired load distribution?

If we use a conventional form of connecting rod as shown by a broken line at D in Figure 10.3, then the 'bed' BCB will be relatively soft at B and hard at C, giving a heavy concentration of load near C which will limit the overall capacity. The distribution will be somewhat as shown by the broken line P. It is difficult to see how we can fix anything to the big end without producing a hard spot and corresponding local loading.

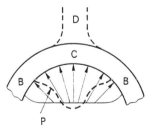

Figure 10.3 Desirable loading on big end from crank pin

The answer is a simple one when we recognise it. Any strut, like the shank of the connecting rod, presented at right angles to the ring-beam of the big end, will cause a local deflection. The least effect will be caused by a *tangential* strut, but we need the resultant force in the supports to go through the centre of the crank pin. The answer is to provide *two* such struts, as in Figure 10.4, so that the section of big end A to B which has to

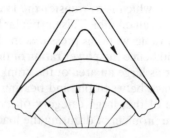

Figure 10.4 Twin strut connecting rod

carry the load works like a suspension bridge or a hammock. The principle of diffusing a load into a thin weak shell by tangential members was seen also in balloons which suspended the basket from a netting, and in the cryogenic rail tank of Section 6.5.

Figure 10.5 shows a connecting rod by Wasa, the Finnish company which introduced this design. By stress analysis they calculated that it would increase the effective bearing area and hence the load capacity and the rating to which the engine could be raised by 40 per cent. They carried out

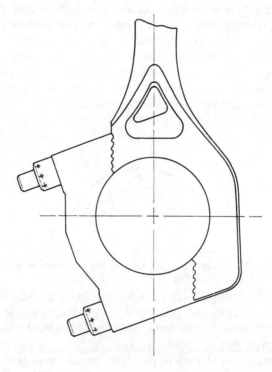

Figure 10.5 Wasa engine big end

tests on actual engines which confirmed the calculated improvement and they then adopted the design.

In the fairy story a queen wanted to be sure a prospective daughter-in-law was a true princess, so she put a pea under her mattress. The girl complained in the morning of the lumpy bed, and a second mattess was added, and so on. Even under thirteen mattresses the pea caused the girl to sleep badly, and the queen concluded that she must be a genuine princess.

The conventional connecting rod creates a hard spot in the relatively flexible ring-beam of the big end, and the Wasa design cures it in an elegant and satisfying fashion. The moral of the fairy tale is a bit dubious, but the moral of the Wasa connecting rod for the designer is quite clear. Strive for clear insight into the way components work as structures and choose where possible good forms which function in clearly understandable and straightforward ways. The 'twin strut' connecting rod exhibits clarity of function very convincingly.

Further thoughts

Connecting rods have been important engineering components since the introduction of the rotative steam engine, but new thoughts on them are still possible. At the present time, several groups are looking at filament wound rods, where most of the strength lies in strong fibres. It is difficult to manage the details, but the incentive lies in the reduction of weight which might be achieved and would improve the smoothness of running.

An interesting thought stems from the twin strut connecting rod, and the view of the most heavily loaded part of the big end as a kind of hammock for the crank pin. To what extent might it be desirable to increase the flexibility of the big end so as to make it conform well to the crank pin? In the extreme case, imagine the big end very flexible, a thin strap. Then it will tend to wrap itself closely round the crank pin, and it will be like a belt, a frictionless one because of the oil film, exerting a uniform pressure on the film, which we might expect would adapt its thickness to suit. Work has been done on this very elegant idea, the foil bearing, and it has found application for some special purposes. It is not suitable for a connecting rod, however.

10.2 A SUITCASE HANDLE AND A SUSPENSION ARM

Many engineering failures are the result of poor structural design, and this section discusses two with similar errors but different consequences.

Setting out on a journey some years ago with a new suitcase, the writer was disconcerted when it fell off its handle. The essential parts are shown in Figure 10.6a.

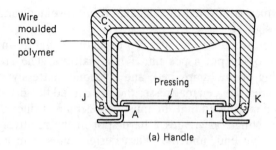

(a) Handle

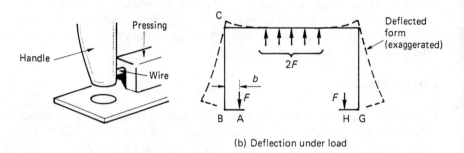

(b) Deflection under load

Figure 10.6 Suitcase handle

The handle did not fail because it was not strong enough, but because it was not stiff enough. Consider the simplified representation in Figure 10.6b, where ABC etc. is the handle. The whole section BC was subject to the bending moment Fh, where F is half the suitcase weight and h is the length of the overhang AB, and bent as shown by the dotted line. Moreover, the bending increased the length of the overhang h, increasing the bending moment in the lower part of BC, so increasing h still further, and so on. It is true that friction would oppose this motion, slip between handle and pressing causing the force F to have an inward component, but a designer would be rash to rely on such an effect: in this case, the handle failed.

It is difficult to believe that the manufacturer had done any testing, for the weight in the case was not abnormally high. It is possible that the overhang h was larger than average in this particular sample. The tolerances in such a product would normally be wide, and there may have been above-average end play between the polymer and the pressing.

The failure revealed another weakness in the design. Once the handle was off, there was nothing to enable a substitute to be contrived, nothing to which anything could be attached. The writer saw other suitcases afterwards with similar bodies, but none with the same handle.

When an attempt was made to lift the case, the handle simply disengaged. It could easily be replaced, but would not support the weight of the case. However, by making notches with a knife at J and K (Figure 10.6a) in the polymer, winding some string round the notches and tightening the string by frapping (pulling the string together sideways with a few additional turns), the handle was retained in place and has lasted for years. The sides of the handle start to bend out as before, but this induces a load in the string which reduces the bending moment everywhere above J and K. This is a redundant structure, the load in the string being determined by the flexibility of the parts. This is at variance with the principle of least constraint, but is an effective solution.

What could be done in a redesign? Simply bending the very ends of the wire upwards would cause them to hook behind the edges of the holes in the pressing, so that the pressing itself would perform the same function as the string, that of preventing A and H moving outwards.

The suspension arm

Figure 10.7a shows the suspension arm, which is made from a steel pressing. It is pivoted to the wheel carrier at A and to the car along XX, while a coil spring acts on it at S. There are thus upward forces at A and XX on the arm, those at XX totalling about one-quarter of the weight of the car. The attachment at XX is made via rubber bushes in steel housings fitting into two holes H in the arm, which are pressed out with deep flanged rims, so that the forces F applied to the arm act in a plane offset a distance h from the plane of the web of the ends of the arm, as in Figure 10.7b.

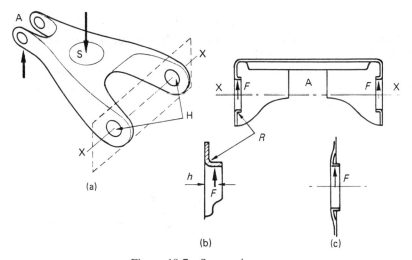

Figure 10.7 Suspension arm

Just as in the suitcase handle, the force F acting on the arm h produces a bending moment in the vertical member, here the sheet metal web of the arm, but in this case the trouble arose from the stress, not the deflection. Fatigue cracks originated close to the flanged holes, and the part had to be modified.

The designer originally provided a generous radius between the flange and the web, at R in Figure 10.7b, with the object of avoiding a 'stress-raiser'. However, increasing the radius increased the overhang h, which was a bigger effect than any stress concentration factor. The arm was redesigned with a much smaller radius and a reduced overhang h.

Notice that a more elegant version would be the form shown in Figure 10.7c, with a conical 'joggle' in the web in order to put the force F into the plane of the rest of the web (but it would be necessary to check that this did not lead to difficulties in pressing).

The suitcase handle and the suspension arm present very similar characteristics, and indeed are both rather like the kind of rotor considered in Section 3.8. In design, similar situations keep recurring in different circumstances, and the designer learns to recognise them, and has a stock of useful ideas on tap to deal with the problems they present.

10.3 A WINDOW STAY

Casement and fanlight windows are usually fitted with a stay to enable them to be held open to varying angles, as in Figure 10.8. At first sight, such a fitting would not seem to present much of a design problem, but it will be seen what a large number of considerations there are and how much thought is needed.

To begin with, let us study the version shown in Figure 10.8. The stay is pivoted to the window at A by a joint which has two degrees of freedom, in the horizontal and vertical planes. It is like a Hooke's joint, except that the two axes do not usually intersect. There are holes in the stay, and two pegs B and C fixed to the window frame. When the window is open, one of a number of holes in the stay engages the peg B, and when the window is

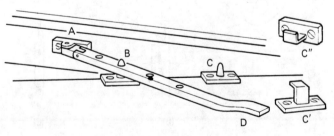

Figure 10.8 Window stay

closed both B and C engage holes, preventing the window being opened from outside. The peg C is often replaced by an outward-facing hook (C'), the advantages of which will be explained later. Alternatively, the fixing C may be attached to the window, like A (C").

The most popular material is aluminium alloy: brass or bronze is attractive to many people but expensive, and steel with some protective treatment is a possibility. The discussion will be based on aluminium alloy.

Should the part C be attached to the window or the frame? One criterion on which to base this decision might be, which will better resist forces tending to open the closed window, whether from a prospective intruder or wind suction? The answer is, the location C" on the window is stronger: if the ratio AB:BC is 1:3, and if a force F is applied to the window, then with a hook or peg at C on the frame, the loads will be F at A, $1.33F$ at B and $-0.33F$ at C.

With a hook at C" on the window, there will be a load $0.75F$ at A, F at B and $0.25F$ at C". So far, this looks like a strong argument in favour of fitting C on the window. Figure 10.9 shows the loading system for this arrangement, which will be adopted henceforth, and the bending moment diagram, which shows the maximum value of $0.75FL$ opposite the peg B.

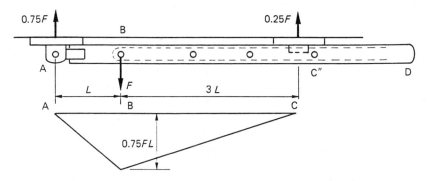

Figure 10.9 Bending moments in window stay

The main body of the stay is generally of a channel form with the open side of the channel facing down. This section provides good strength in bending, and also engages the top of the peg B when resetting the stay to a different postion, which is useful, as we shall see. The flanges are usually tapered away at D, the free end, which is also joggled upward to provide a convenient raised handle for operating the stay (see Figure 10.8). The smooth upward face of the channel looks neat and does not collect dust.

Now at this stage it might be thought that the stay could readily be made from a length of aluminium alloy extrusion, with the web of the channel section cut away near A to provide the lugs or trunnions for the first stage of the joint there (Figure 10.10). This solution would be cheap and profit from the superior qualities of the wrought material.

However, we shall see that the structural requirements make the uniform basic section of the extrusion unsuitable, and the numerous small operations needed put up the original, attractively low, cost. Extrusions are an excellent resource for the designer, as was seen in the quarter-turn actuator, but may not be right here.

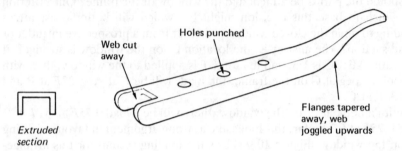

Figure 10.10 Stay formed from extrusion

The bending of beams with cut-outs in the web

Where a bending moment diagram such as Figure 10.9 applies, it is easy to concentrate the attention on the region where the moment is a maximum, near B in this case. Sometimes, however, and the window stay is an example, there are weaknesses near the ends and the trouble occurs there (the 'oil-canning' rotor end of Figure 3.16 was another case).

Figure 10.11a shows the joint at A in one possible form. The pins at A and E provide the two necessary degrees of freedom. A ball joint would also serve, and it is an interesting exercise to consider if there is any form in which it might be competitive.

Consider the transfer of the load $0.75F$ at A into the stay, using the plan view in Figure 10.11b. At E there is a bending moment $0.75Fh$, where h is the perpendicular distance between the axes A and E, which will give rise to forces P in the flanges, where P is $0.75Fh/d$ and d is the distance between the centres of the flanges (see Figure 10.11c). The shear force of $0.75F$, however, falls on the window-side flange, which is therefore subject to a bending moment $0.75Fe$ at the point G (see Figure 10.11c).

The arm e is small compared with AB, but the solitary flange at G is shallow compared witht the depth d, and it is probable that the designer would draw the flanges too thin at this point unless he considered this effect specifically. It would also be extravagant of material to provide the thickness of flange needed at G throughout the length of the stay, which is a reason against using an extrusion.

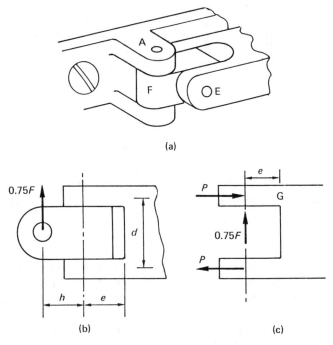

Figure 10.11 Joint in stay

It is important to recognise that where a cut-out removes the web of a beam there may be bending in the flanges and great weakness. The planet carrier of Chapter 8 could be regarded as a ring-beam with cut-outs and the suspension bridge tower as a straight beam with cut-outs. Notice that it is the shear force that causes the weakness, however, the bending moment in a flange being equal to one-quarter the shear force times the length of the cut-out, except in a case like that of the window stay where only one flange contributes.

An ergonomic aspect

Window stays and shampoo bottles have this in common, that we frequently use them when temporarily or partly deprived of vision, by darkness, relative position, or fear of soap in the eyes. Consequently, shampoo bottles should be stable on their bases and window stays should be easy to engage without being able to see the holes and the pin.

However, some window stays are hard to engage by feel because positions in two degrees of freedom have to be correctly matched for the pin to go in the hole. Thus with the section of Figure 10.12a, it is possible for the pin to go past the hole as we push the stay in or out, even though we

202   FORM, STRUCTURE AND MECHANISM

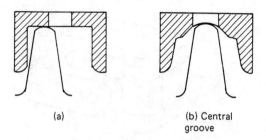

(a)　　　　　　　　(b) Central groove

Figure 10.12　Finding the hole

press down on it. With the section of Figure 10.12b, the central groove removes one degree of freedom and the pin goes in the hole every time.

Other aspects

Many more points could be made about even such a simple device as a window stay. Here is one. Can we make a stay which only lets the window open a little way, but can in an emergency, such as a fire, be freed completely *only from the inside*?

The key point to recognise is, that the stay can be reached from outside almost as well as from the inside *except when the window is fully closed*. Figure 10.13 shows a solution. The pin has a shoulder and a large head. It passes through a slot in the stay, and this slot is enlarged locally at several points to a hole which will admit the thicker part C of the pin below the shoulder, so fixing the stay. When the window is fully closed, the pin comes

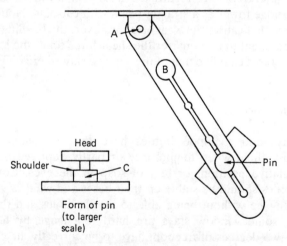

Figure 10.13　A secure stay

opposite the only hole (B) in the stay which is bigger than the head of the pin, so that the stay can be disengaged.

The reader might like to think about the possible configurations of the joint at A. Is the nesting order the best or should the horizontal axis E be nearer the window and the vertical axis A further from it? In the joint as shown in Figure 10.11, should the axis E be on the opposite side of A to the handle? There are designs like that, with the joint A overhung, not straddled.

The use of a hook rather than a pin at C was mentioned earlier. If such a stay is carefully fitted, when the window is closed, the stay is pushed down under the overhang of the hook and springs back up, leaving a measure of prestress in the system which prevents rattling. This raises some new questions. What can be done by the designer to make the stay easier to fit well? Should some large deliberate spring element be introduced?

10.4 THE *CHALLENGER* DISASTER

In January 1986 the space shuttle *Challenger* exploded shortly after lift off, with the loss of the seven people on board. The cause was established to have been the failure of the seals of a joint in the tubular steel case of one of the two solid fuel boosters, which allowed hot gases to escape and weaken adjacent structural parts.

The case was a steel tube roughly 3.6 m in diameter and 12 mm in wall thickness, made of segments about 4 m long jointed together. It was the seal on one of the joints which failed: a section of the parts involved is shown in Figure 10.14a.

The lower rim T of the segment above the joint slides into a circumferential slot S in the upper rim of the segment below and the two are secured together by 177 radial pins. The pins are 25 mm in diameter, and a quick calculation shows that their pitch is about

$$3600\pi/177 = 64 \text{ mm}$$

which is in a suitable proportion to their diameter at rather more than twice. This arrangement is called a clevis joint, a traditional term for a common joint for rod ends which consists of a blade or tang fitting into a fork or clevis (like cleft, split) and secured with a single pin.

The sealing arrangements consisted of two O-rings as shown in the figure. These lost their elasticity at the low temperatures prevailing on the day of the disaster, so that they were unable to expand sufficiently to follow the tang surface when it moved away from them, which it did for the reasons that follow.

204 FORM, STRUCTURE AND MECHANISM

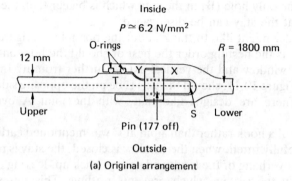

(a) Original arrangement

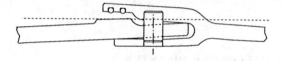

(b) Deflections when firing

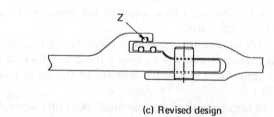

(c) Revised design

Figure 10.14 Joint in shuttle booster case

The firing of the solid fuel pressurised the case to about 6.2 N/mm², producing a hoop stress in the 12 mm wall of about

$$\frac{6.2 \times 1800}{12} = 930 \text{ N/mm}^2$$

which would cause a radial stretch of the order of

$$\frac{930 \times 1800}{200,000} = 8.4 \text{ mm}$$

and away from the joints the case would indeed expand by this order of movement, slightly reduced by longitudinal stresses and Poisson's ratio effects. However, at the joint, the effective thickness of the material was about two and a half times as great, so the unconstrained stretch would be much less, about 3 mm. The effect is to pull the casing inwards locally, just as an elastic band round a sausage balloon would pull it in locally. The effect, exaggerated for clarity, is shown in Figure 10.14b. The bending inwards opens a gap opposite the O-rings, estimated at 0.74 mm at the inner one (the nearest to the high-pressure region) and 0.43 mm at the outer one. The reader may like to consider this design and how it might be improved.

Apart from the unsatisfactory choice of material for the O-rings, what is wrong with the design? The choice of a clevis joint may seem strange, but what are the alternatives?

Let us start by considering the functions of the joint, which are two, the structural one of resisting the loads and the sealing one, or preventing bursting and preventing leaking.

Resisting bursting

The prevention of bursting requires a strong joint which takes into account the nature of the case segments, which are very thin for their diameter and hence quite flexible. We should apply the principle of short direct force paths, and particularly, we should aim at local closure of the force loop. A screwed joint between two such segments, apart from the difficulty of screwing it together, would fail under quite a small tension, the radial forces stretching the outer part and compressing the inner part until the threads disengaged.

A screwed clevis joint, as in Figure 10.15, would meet the local closure principle, but a simple calculation will show it would have to be very thick in the region A to withstand the bending. It is a much less attractive solution than the 'pin-and clevis' joint, even if it could be made and assembled easily.

Could the clevis joint have been designed to be elastically matched to the rest of the case, so that under pressure it would stretch the same amount? The reader is invited to think about this. The nature of the pinned joint means that there must be extra metal. This could be made flexible by

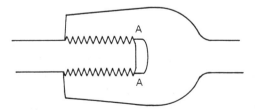

Figure 10.15 A screwed clevis joint

making a circumferential corrugation or wave in the clevis, so that under pressure it would extend more easily. Unfortunately, apart from the fearful problem of manufacture, the stresses produced are too high unless both the local thickness and the amplitude of the wave are made very large (seven times the general wall thickness of the case if the wave is sinusoidal).

The reader might well ask why the flanged joint discussed in Section 3.7 would not suit. It would certainly fill the function of resisting bursting, but it would be bulky and heavy compared with the clevis joint. In Figure 3.13, the additional cross-section of material in the flanges is $15t$ by $16t$, or $240t^2$. In the shuttle booster case joint the additional thickness is about $1.4t$ over a length of about $7.5t$, or about $11t^2$. The flanged joint could be pared down a lot, especially by reducing spanner clearances, but not to anything like as little as $11t^2$, and in a space shuttle, mass, and to some extent, overall diameter, are at a premium.

Having established the virtues of the clevis joint, it is time to examine the prevention of leaking, the function in which the joint failed.

Sealing

In other cases where preventing bursting and preventing leaking were needed, it helped to separate the two functions. Figure 10.16 shows in principle how this might be done. A tubular piece B spans the joint at J (of unspecified kind, but on the basis of the argument so far, a clevis joint), and seals on the casing at S and T.

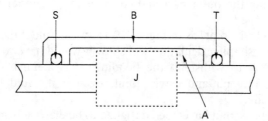

Figure 10.16   Separation of functions

The part B would have to withstand the full internal pressure except at S and T, since the space A would be at atmospheric pressure, and this fact could be used to ensure that it matched the expansion of the casing. It would be essential to ensure that it did follow the casing when the internal pressure first rose (there is a long discussion in the report of the Presidential Commission dealing with this problem as it affected the O-rings), but essentially this solution could be made to work. However, the problem is not so difficult technically as to make this bold solution necessary, and it would be bulky, and awkward to accommodate without extensive redesign.

What was wrong with the original design? The elastic mismatch between the rest of the case and the thickened belt formed by the joint caused the parts of the two segments to separate opposite the O-rings. Could the O-rings be moved to a place where the gap remains the same or decreases on pressurisation, such as X in Figure 10.14a? The answer is no, because of the leakage that would occur past the clevis pins: the O-rings must be placed earlier in the leakage path than the pins.

The solution adopted is shown in Figure 10.14c. The extra tang on the inside diameter of the upper segment contributes nothing to the prevention of bursting, but it provides a location, at Z, which is right upstream in the leakage path and at which the gap closes on pressurising. It is not elegant, but it is effective and adds little to the bulk. Some other internal changes were made for added safety.

The reader may have spotted a very simple change which might work, the insertion of a thinned section at Y in Figure 10.14a. This would provide a flexural hinge between the rest of the structure and the part with the O-rings, which would be forced outward against the mating part by the internal pressure, so ensuring a seal. The problem arises in such a solution of ensuring that it actually seals on the first rise of pressure. Also, it is bulkier than might be expected.

Because the segments are recovered after the launch of the shuttle and used to build more boosters, the flexural hinge at Y may have to be used several times, and so it cannot be allowed to yield too much. This limits the design stress, which in turn demands a fair length in which the bending can take place. Unfortunately, this additional length raises the bending stress due to the internal pressure, and so on: this sort of difficulty is common with flexural elements and occurs in the diaphragm used to enable the planet pinions to self-align in Section 8.5.

Pedestrian though it may be, the solution of Figure 10.14c seems to be the best answer.

11

The Principles of Design

11.1 AN EMERGING DISCIPLINE

Functional design does not show the ordered structure of concepts and principles characteristic of a subject like physics, partly because it has not been systematically studied, but partly because it is not a reflection of a nature that appears to have immutable laws. Principles are emerging, however, which can provide valuable guidance to the designer, and several of them have been mentioned in the course of this book. In this final chapter they are drawn together and surveyed.

A great problem in discussing design is the complexity of the relationship between aspects, which makes it difficult to describe in a serial fashion. A parallel is the description of the layout of a town. As a visitor wanders around, he will keep seeing a conspicuous building or come upon a particular street at different points in its length. When he comes to a crossroads, he can only take one road away from it, leaving the others till later. In the same way, an account of the underlying principles of design, or even the rationale of one example, cannot be arranged as a simple unbranched chain of argument.

Engineering designers follow principles in their work, even though they may not be able to set them down in words: many of them will recognise some parts or aspects of those given here. These principles may often be ones they have developed themselves in the course of their work, in some specific form or application. The principles of Least Constraint and Kinematic Design are probably the most widely known. Many others are not covered in this book, some few no doubt because they are unknown to the writer, many more, especially those to do with manufacture, for lack of space. Communications about any such omissions would be gratefully received.

11.2 LEAST CONSTRAINT AND KINEMATIC DESIGN

The Principle of Least Constraint was given in Section 4.3, in the form:

in fixing or guiding one body relative to another, use the minimum of constraints.

It has been evoked in many examples, but particularly in Chapter 9, in respect of hydraulic pumps.

The idea of kinematic design is rather more extensive, in the writer's usage, at least. It might be stated, very loosely, as 'if you can, let it go free'. It embraces the important principle of introducing an additional degree of freedom to balance out loads, as was seen in epicyclic gearing in Sections 4.4 and 8.5, and in the tilting pad bearing in Section 4.7. A classical example is the differential, which introduces an additional degree of freedom between the ends of an axle, to balance the torques on the two wheels. Another aspect of kinematic design is provision for relative movement due to strains, as in the discussion of a centering method in Section 1.4 or the conical washer in Section 3.5.

Elastic design

Complementary to kinematic design is the concept of elastic design. If the principle of kinematic design can be expressed as 'avoid fights between components', elastic design can often be reduced to 'if there is going to be a fight, ensure the loser gives in easily and no-one is hurt'. The use of flexural joints is an example (Section 5.11): a mechanism is designed which, regarded as a complex of rigid bodies, has insufficient degrees of freedom, so that for it to work something must bend. That something must bend sufficiently without developing stresses it cannot sustain.

Roller and similar bearings are another case, not reducible to the 'fight' analogy, where kinematic design would lead to only three rollers. It is more practical to accept the redundancy and make the parts very accurately so that the load sharing is satisfactory.

Yet another case is the use of quill-shafts, very flexible in torsion, to link the layshaft gears in reduction gears of the form shown in Figure 11.1 Because these twist a lot under the torque, reasonable load-sharing between the two layshafts can be achieved with inexpensive tolerances on the parts. The gears A and B are carried on their own integral hollow shafts in their own bearings, and are only connected in torsion by the quill-shafts. This could, of course, be seen as an example of separation of functions, side loads being catered for separately from torque.

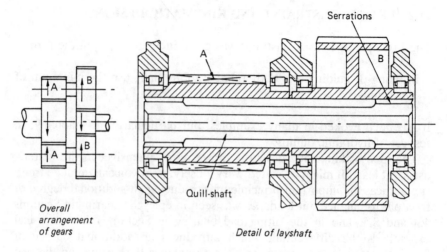

Figure 11.1 Twin layshaft reduction gear

11.3 SEPARATION AND COMBINATION OF FUNCTIONS

Separation of functions is usually of benefit when the functions concerned are intrinsically difficult or important or involve conflicting requirements. In Watt's invention of the separate condenser there was a conflict of a very fundamental kind – the working stroke required a hot environment, the condensation a cool one (Section 1.8).

In the common case of separating the functions of resisting internal pressure and containing fluid contents (preventing bursting and preventing leaking), the first function involves high stresses and hence deformations, while preventing leaking at running seals demands dimensional stability. This common theme of dividing the functions of preventing leaking and preventing bursting has been illustrated by the cases of a gear pump (Section 5.6) and the cylinder liners of the Clerk hydraulic machines (Section 9.7).

As has been explained, the twin layshaft reduction gear shown in Figure 11.1 can also be regarded as a case of separation of function: the transverse support of the layshaft gears is required to be stiff, whereas the torsional link between the two, the quill-shaft, should be as flexible as possible, so separate provision is made for each function.

Combination of functions is usually aimed at reducing costs, and is not common in products of high performance and exacting specifications. An example was given however, in Section 1.8, the trailing arm rear suspension with the two arms linked by an internal beam of high bending stiffness and low torsional stiffness.

11.4 DESIGN MATING SURFACES OR ABUTMENTS

All aspects of a product must be defined, and to that extent, designed, before it can be manufactured, but abutments deserve special attention. Moreover, the very act of concentrating on them can be a real help to a designer in the early stages, when making a start can be difficult. Examples given were the mounting of a bush in a casing, big end joint faces and a gas turbine stator blade mounting (all in Chapter 3). In Chapter 5, the relief of mating surfaces to reduce 'sticky drawer' effect was mentioned, and the bevel gear mounting in the same chapter is also an example of attention to mating surfaces.

In general, as soon as a design reaches the stage where it is beginning to be thought of as an assembly of separate parts, careful thought should be given to the abutments of those parts, to see that they conform with the principles of kinematic design and clarity of function, that they can be manufactured readily to whatever accuracy is required, and that they are not overstressed nor yet too large. Above all, they should be in the right places, which means, very often, where the loads on them are not too high and their accuracy is not made unduly important by poor kinematics, for example, too short a base-line.

Attention should also be given to the number of setting-up operations required in the manufacture, which should be one if possible but may sometimes have to be two or even more: extra operations are not only costly in themselves, but tend to reduce accuracy, particularly of squareness and alignment.

11.5 CLARITY OF FUNCTION

This principle has been left to this point in order to have covered first some of the many others it includes within its scope. For indeed, this is an all-embracing idea, a general philosophy, rather than an aid in particular design problems.

The principle of clarity of function may be stated as follows: every function in a design should be achieved in a clear, simple way, as directly as possible and without ambiguity.

To some extent, this principle is reflected in the concept of economy of means: to use both belt and braces to support trousers would not exhibit clarity of function, though it might reassure the very nervous. But clarity of function is easier to recognise than to define, and it can be recognised in some of the related principles.

For instance, the Principle of Least Constraint might be regarded as requiring clarity of function in the supporting or guiding of one body relative to another, with just sufficient constraint being provided.

The principle of uniformity

This is another principle belonging to the 'clarity of function' stable. We should aim at uniformity in performing functions.

For instance, if we use a number of fasteners to join two flanges, we naturally make them all the same size, not three big and three small, say. If we design a tie, we make its cross-section the same from end to end. These examples are trivial, but more substantial cases arise in structures, for instance, in the design of cylinder heads (Section 5.4) where the aim is to approximate to uniformity in a complex casting full of exhaust and inlet valve ports and similar features which produce a host of intersection problems. For example, in such a structure, the designer will aim at spacing the bolts or studs as regularly as possible around the circumference of the cylinder.

It is worth noting, however, that deliberate departures from uniformity may be made for special purposes, for example, a slight departure from regular spacing of the bolt holes in two mating flanges to ensure correct assembly: a dowel with an easy fit in one flange and a tight one in the other can also be used.

In castings, it is desirable to keep the general thickness uniform to avoid voids or sinks, but here, as elsewhere, uniformity is a good discipline for the designer and also has an aesthetic value. Uniform thickness in outside webs and uniform fillet radii give a pleasing effect of unity.

The discrete joint

In Section 3.3, in connection with the big end, the principle was introduced that a joint should be designed so that it should behave as nearly as possible as if there were no joint. This is an example of clarity of function, the ring beam of the big end being made as nearly continuous as possible.

In the flanged pipe joint (Section 3.7), all that the flanges and bolts do is to force the two sections of pipe wall hard together, and all the loop followed by the force paths, out through one flange, through the bolts and back through the other flange, is just a tiresome diversion needed for assembly or manufacture, or perhaps, for transport.

This line of thought leads easily to the joint form of Figure 11.2, where the rings R are welded or otherwise rigidly fixed to the pipe end and the ring S, made in two or more parts, is clamped round the circumference. Notice how the rings R do not touch one another and the normals N at the clamping surfaces pass through the centre of the abutment of the pipe ends at E. These last two features demonstrate clarity of function.

The separation of functions may be regarded as an instance of this same general principle of clarity, as may be the design of abutments, where we decide just how to fit parts together in a strong, economical and easily

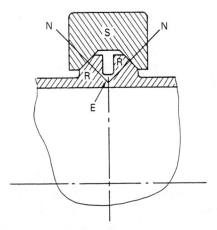

Figure 11.2 Clamp type of pipe joint

manufactured way. The principle is very wide and general, amounting almost to an outlook or design philosophy: only occasionally is it of direct help in itself, rather, it is a criterion which we can use to test our solutions. Does this design show clarity of function? and if not, why not, and what can we do about it?

11.6 SHORT DIRECT FORCE-PATHS

This is yet another principle which could be considered to be an aspect of clarity of function. It was introduced in Section 5.3, and has not figured much in the text. Indeed, the connecting rod described in Section 10.1 might be felt to contravene the principle, and certainly shows how carefully it needs to be applied. The epicyclic planet carrier in Chapter 8 is an example of the high penalty in weight and cost which may be imposed when, because of intersection problems, we are forced into devious force-paths. In such cases, it is sensible to look for radical or Alexandrian solutions (Section 8.5)

Local closure

An important corollary of the principle of short direct force-paths is that of local closure, the idea discussed in connection with the *Challenger* field joint (Section 10.4). A bolted flange joint meets this criterion, the loop being closed from pipe abutment, through flange, bolt, and second flange, and so back to the pipe abutment again: a portion of the circumference, sawn out like a slice of cake and containing one bolt, is a joint in itself. The joint of Figure 11.2, on the other hand, relies on the circumferential band S for its integrity, and a section cut out would fall apart.

As an example, the reader may like to think how the joint form of Figure 11.2 might be converted to local closure. It is easily done, if a seal is not required, but there is an intersection problem which makes sealing very difficult.

11.7 MATCHING AND DISPOSITION

These are recurring problems in functional design (Section 1.7), which may generally be tackled in quite a systematic fashion. Matching was encountered in the springs of the bathroom scales (Section 2.9), where the diamond-shaped cut-out matched the section modulus to the bending moment, in the inclination of the supporting straps in the cryogenic rail tanker (Section 6.5), and in the biasing of the web of the diaphragm in the epicyclic gear planets in Section 8.5. A difficult, perhaps an insoluble, problem of matching was seen in the field joint of the shuttle booster in Section 10.4. In the case of the nut and bolt, it was shown that due to a happy accident, the matching of the threads under load was good, giving a fairly uniform distribution of load (Section 3.6).

Disposition

Disposition, the sharing out of some resource in short supply, like length, or area, or strength, to the best effect, has been described in the case of the cassette spool (Section 1.7), the fir-tree root (Section 3.6), and the quarter-turn actuator (Chapter 7), three cases where the resource in question is a length, and in the big end joint, where the resource is an area (Section 3.3). One more example is given here, from large alternators. The rotor in these large machines, of the kind used in power stations, is simply a huge electromagnet, constantly excited, with two poles. As it rotates, the field sweeps through the surrounding stator coils, generating electricity. The rotor is of steel, and the winding is of copper bars in longitudinal slots, as in Figure 11.3a.

In this case, the resource or commodity in short supply is the area of the circular cross-section. The diameter is limited by the speed of the machine, which is fixed by the need to generate at 50 Hz or 60 Hz to 3000 or 3600 rev/min, and the centrifugal loads. The object, given this more or less fixed size of circle, is to pack into it the most powerful electromagnet possible without too high losses. Doing so will maximise the power per metre length of the alternator, a highly desirable thing. The designer should therefore make sure the best use is made of every scrap of area of cross section. Consider the functions to be provided in that area.

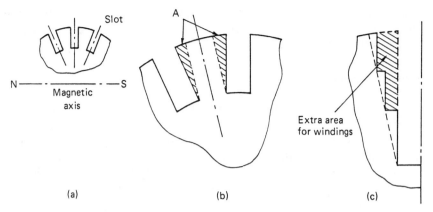

Figure 11.3 Alternator rotor slot forms

The steel rotor provides the path for magnetic flux and the strength to hold the whole together against centrifugal loads.

The slots contain the windings, which provide the path for the magnetising current. The cross-sectional area of the slot is therefore filled with copper conductors, the insulation for them, and cooling passages (the coolant is usually hydrogen). It looks as if every bit of cross-section is contributing.

However, up to the mid 1960s the slots were generally parallel-sided, as shown by the solid line in Figure 11.3b, so that the steel teeth between them were thinner at the root than at the tip. Now the centrifugal load is clearly greatest at the root, so that the extra width towards the tip makes no useful contribution to strength. Also, the narrow root forms a bottleneck for magnetic flux, so that the extra width does not contribute usefully to the flux path. In fact, the wedge-shaped areas A in Figure 11.3b are wasted: they should be used to provide room for more conductor, insulation and coolant passage. The teeth should be parallel-sided, and the slots tapered.

Now one objection to this change advanced by a company to which the writer suggested it, was that the conductors would need to come in several sizes. These conductors are shaped copper bars with cooling channels in them, not simple wires, packed in about eight layers, usually two in each layer. Eight sizes would be needed with a tapered slot, since the width would be different at each layer. However, if the slot is made with steps, say, three steps, as in Figure 11.3c, then only a few sizes would be needed. With three equal steps, two-thirds of the wasted area can be recovered, with four, three-quarters, and so on. (These fractions are not exact unless the number of layers is divisible by the number of steps: the recovery with three steps and eight layers is only 65.6 per cent, instead of 66.7 per cent.)

These fractions show the rough rule mentioned in connection with the fir-tree root (Section 3.6), that the unrecovered part is inversely proportional to the number of steps.

Cutting the cake

An interesting aspect of this case was the use to which the extra area of conductors was put. The increase was 16 per cent, but the magnetising current was only increased by 8 per cent, so that the current density was decreased, the ohmic drop was decreased, and the efficiency improved. Thus half the extra area was used to increase output per metre and half to increase efficiency, but other trade-offs would be possible, according to the economic factors involved.

A designer should always study carefully how to share out any such 'cake' to the best advantage. In the more common case of kicks to be distributed, it is also important to distribute them so as to produce the least overall harm.

11.8 NESTING ORDER AND RELATED PRINCIPLES

Nesting orders were discussed in Sections 5.7–5.10, and examples were given from typewriters and printers, machine tools and vehicle suspensions. Some principles were given for choosing orders, including:

lightest, smallest, fastest, most frequent motions should be furthest from earth,

and the converse. These are only guides: it is easy to find cases where there are good reasons for departing from them, and sometimes, of course, the lightest may happen to be the largest (usually, in such a case, lightness will take precedence).

Another principle is that branches should be short, to keep down the complexities which arise from stacking motions. It is desirable, also, to avoid building branches on difficult motions, like continuous rotation. An example is given by current attempts to devise practical ways of varying the valve timing of petrol engines with the speed.

Because the cam shaft is turning at high speed through an indefinitely large angle, it is undesirable to stack any other motion inside or outside its rotation. It is practicable to move the camshaft axially, and make the cam form vary along its length, but this is not attractive because it becomes difficult to accommodate everything lengthwise (there is a disposition problem). Accordingly, the possibility of using another branch from earth should be studied, and Jaguar have invented an elegant way of using the tappet.

Figure 11.4a shows a conventional arrangement of a cam with a flat-faced tappet or follower which is urged against the cam by a force F from a spring. The tappet reciprocates along the axis XX under the action of the cam as it rotates. Jaguar propose to make the tappet capable of rotation about the axis XX, and instead of it having a flat face, to provide it with one of the form shown in Figure 11.4b, a shallow valley with sides sloping up at an angle α to the flat and the bottom cylindrical, with a radius r equal to the base radius of the cam.

At low speeds, the valley is at right angles to the camshaft, as at (c), so that the action is simply that of a flat faced tappet. At high speeds, the tappet is turned 90° about the axis XX so that the valve will begin to lift α earlier and will close completely α later, in terms of camshaft rotation. Lift and fall occupy the same angles as at low speed, but they are separated by a dwell while the nose of the cam crosses the valley floor, during which dwell the tappet will not move. A perspective view is shown at (e): the rotation of the tappets about their axes is effected by means of gear teeth on their sides, which are engaged by a long rack parallel to the camshaft: moving the rack lengthways turns all the tappets at once.

Notice that the face of the cam must be radiused across its width to the radius r, so that at low speeds the cam will fit the valley floor (see Figure 11.4f, which is a view at right angles to the camshaft at low speeds).

The beauty of this invention lies in the use of the movement of the tappet, which is a spare unused branch, as it were, rather than some complexity applied to the camshaft, which starts full of difficulty because of its high speed of continuous rotation.

Related to these ideas of nesting is the principle of minimising the inertia of moving parts, of which reducing the unsprung mass is a particular case. The guides to nesting order are only that, guides, and sometimes reducing inertial loads is the reason for departing from them.

11.9 AVOIDING ARBITRARY DECISIONS: COMBINING GOOD FEATURES

There are many other principles which are of use in design, and many, like the thermodynamic ones, are not of relevance in this book. There are two broad ones, however, of very wide application: one is the principle of avoiding arbitrary decisions. This is a counsel of perfection. It is impossible in practice to get started at all without making arbitrary decisions, but it is important to be conscious they have been made and to go back and review them as the work develops.

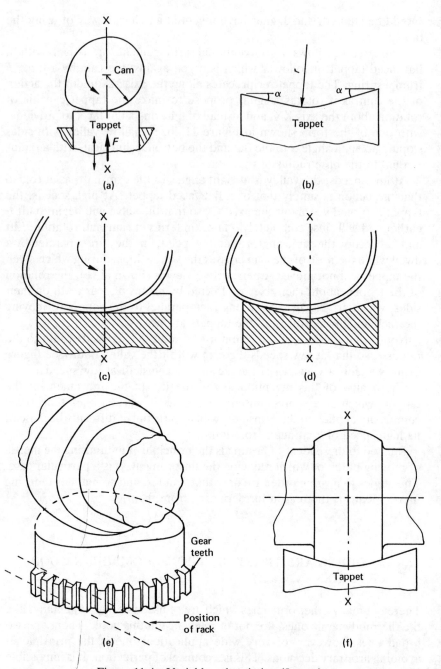

Figure 11.4 Variable valve-timing (Jaguar)

Another principle, a very obvious one, is to try to combine good features in one design. Where two different constructions have distinct advantages, try to find a third which combines the merits of both. The way forward may prove to be a hybrid of the two original constructions, or it may involve a quite different approach. However, there is a risk that designers, particularly the most talented designers, will overreach themselves in this attempt to have one's cake and eat it. When such a notable goal of combining virtues appears almost within reach, judgement may be clouded and constructions adopted which involve unwarranted technical risks.

These few valuable principles encapsulate much of the content of this book. Simple as they appear, they are not easy to follow, and the writer still sometimes makes mistakes and wastes time through not paying enough attention to them.

Bibliography

GENERAL BOOKS ON ENGINEERING DESIGN

1. French, M. J., *Conceptual Design for Engineers*, 2nd edn, 1985, Design Council Books, London.
2. Pahl, G. and Beitz, W., *Engineering Design*, 1984, Design Council Books, London.

Both these books deal with wider aspects of design than this one. [1] is shorter, more advanced in the application of engineering science to design and wider in the engineering aspects considered. Some ideas are more highly developed in [1]. For example, matching in [2] is confined to cases like the *Challenger* joint, and does not cover cases like the quarter-turn actuator. On the other hand, [2] covers aspects of the management of design which are not in [1].

Because the books have so much in common, particularly the widely-adopted scheme of Figure 1.1 of [1], it should be noted that the date of the first edition of [1] was 1971, that of [2], 1977. The similarities are only such as might be expected between parallel workers in the same field and to the same end, and might be taken to suggest the soundness of both. The reader is recommended to read both.

BOOKS ON MECHANICAL ENGINEERING COMPONENT DESIGN

3. Shigley, J. E. and Mischke, C. R., *Mechanical Engineering Design*, 5th edn, 1989, McGraw-Hill, New York.

4. Juvinall, R. C., *Fundamentals of Machine Component Design*, 1983, Wiley, New York.

These books deal with stress and some other calculations in standard mechanical engineering components. It is assumed most readers have read one of these or a similar book. They give most of the extensive background to setting a design stress, among other things.

OTHER BOOKS

5. Matousek, R., *Engineering Design*, 1963, Blackie, Glasgow.

Matousek's work has many illustrations of detail design and alternative constructions, and is an easy and valuable piece of reading, complementary to this book. Somewhat dated it may be, but to be treasured until something equally good but more modern appears.

6. Polak, P., *Background to Engineering Design*, 1979, Macmillan, London.

The contents of this valuable little book are fairly indicated by the title. It covers much of the same ground as Matousek, but much more besides, on structures, motions, materials etc. Many interesting and useful elements of the design repertoire of means are collected here, with helpful comments.

7. Lewis, W. P. and Samuel, A. E., *Fundamentals of Engineering Design*, 1989, Prentice-Hall, Sydney.

This is the best of many such works known to the author. It combines a general introduction to engineering design, on very light-hearted and readable lines, with a large number of worked examples, showing the application of simple engineering science, mostly stressing, to design problems. Its particular strength lies in the variety and the brisk, clear treatment of these examples.

Index

Abstraction 100, 140
  increasing level of 8–11
Abutment 52–6, 211
Actuator, quarter-turn 144–59
  aspect ratio 148
  calculations 146–8
  crank and slider 147–8
  disposition 152–5
  examples 156–9
  fail safe 158
  guidance 149, 151–2
  Miti-1 157–9
  Norbro 156–8
  rack and pinion 151, 153, 154
  swept volume 145–6
  types of 144–5
Alexandrian solutions 163, 167–8
Alternator 214–16
Aluminium alloy 123–4, 127, 128–9, 142, 150–1
  extrusion 149, 199
Arbitrary decision 7, 192, 217–18
Ashby, M.F. 142
Asymmetry in design 154, 163–4, 212

Backlash 101
Balancing of forces 85, 109, 144, 174, 177, 178, 209
Ball-joint 80, 183
Bathroom scales 38–44
Beam 26–31
  curvature 30
  deflections of 30–1
  human 72
  rationale of 28
  ring 29
Bearing
  ball 36–8
  ball and roller 81–2, 107
  film 86–96
  foil 195
  hybrid 96
  hydrodynamic 86, 91–6
  hydrostatic 86–91
  location 82
  power consumption 88–9
  requirements 111–12
  roller 96
  rolling element 96–9
  squeeze action 91–4
  taper roller 83
  thrust 95–6
  tilting pad 95
  wedge action 94–6
Bending of beam 28, 30, 129
  in carrier 163–4
  in window stay 199, 200–1
  inefficiency of 103–4
Bent-axis pump 180–5
  kinematics 182
Bevel gear 107
Biasing 101–2
  in pinion teeth 173
  of diaphragm 170–1

Bicycle
  tyre 102
  wheel 101
Big end
  joint in 59–62, 75–7, 105
Bird-in-cage toy 9–10
Bolt 56–9, 65–6, 68–9
  choice of size of 70
  function of 59
  stretch in 76–7

Calculations, rough 94
Cassette, camera 15–16
Cast iron 124, 141
Casting 124, 125, 141, 142
Catapult 102
Centering 5–8
Ceramic 138, 141–2
Chain, roller 104
Challenger shuttle 203–7
  seal failure 203
Chamfer 54
Clarity of function 28–9, 42, 76, 168, 179, 195, 211, 212
Clerk, R.C. 188–9
Closure, principle of local 74, 205, 213
Combination 11–12
  of virtues 185, 219
  studying a 150–6
Combinative methods 11–13
Comparator 120–1
Configurations of actuator 144–5
Connecting rod 135, 191–5
Constraints 80–1, 107
  principle of least 83–4, 85
Coupling, face 74
Crank and slider 147–8, 188
Cryogenic tank 134–5
Curvature
  of beam 30
  principal radii of 34
  relative radii of 36
Cut-out 162, 200–1
Cutting tool 138
Cylinder head joint 29, 104–6
Cylinder liner 190

Decision, arbitrary 7, 110
Decomposition 139, 145
Deflection
  of beam 30–1, 38–44
  of springs 41, 44–5

Degrees of freedom 43, 79–81, 84–6, 115, 117, 186–9, 198, 200, 202, 209
Density 128–9
Design 1, 100
  anatomy of 1
  brief 3
  conceptual 4
  diversification in 5
  elastic 209
  embodiment 4
  kinematic 84–6, 179, 190, 209
  stepwise 5–8
Designer 51
Diaphragm 7, 168–71
Differential 209
Disposition, problems of 14–16, 89–90, 119, 152–5, 214–16
Diversification of approach 5
Dowel 73–4
Drive, transfer of 114–18
'Drop-together' construction 179

Ears on can 155–6
Easements 121–2, 184
Efficiency
  of actuator 146
  of joint 64–8
  of use of material 45–6, 130
Electronics 43–4
Ellipsoid 34–6
Embodiment 3
Energy
  in flywheel 48
  in spring 41, 44–6, 133
  in structure 51
Engine
  internal combustion 10–11, 29
  steam 9–10
  Wankel 11
Epicyclic gears 84, 160–74, 186
  freedoms in 84–6
  planet carrier 84, 160–7
  structure 160–7
Essence, seizing the 7, 100

Failure, seriousness of 24
Fatigue 25, 137
Fibre reinforcement 124, 127, 130, 141
Fillet 53
Fir-tree root 66–8

Flange 61
  bend in 166
  of beam 28–30
  of bolted joint 68–71
Flexural elements 118–22, 137, 207, 209
  pivot 121
Flywheel 50
Force path 102–4
Form, relation to manufacture 137
Freedom *see* Degrees of freedom
Functions 11–13, 14–15, 149–50, 215
  combination and separation of 18–22, 210
  separation of 109, 140, 179, 206, 209
Furniture 15

Gas turbine 63, 139
Gasket 29, 77–8, 105
Gear (*see also* Epicyclic gears)
  reduction 209, 210
  sun, planet etc. 84, 160–1
  teeth in actuator 148

Hydraulic diameter 90

Incubation 100
Insight 4, 59, 100, 195
Intersection 16–18, 75, 161

Jaguar 217
Joint
  bolted 68–71, 75–7
  clamp (pipes) 212–13
  clevis 203
  constant velocity 116–17
  cylinder head 104–6
  discrete 212–13
  efficiency 64–8
  general principles of 71–5
  Hooke's 116, 187, 198
  in rotor 72–4
  location of 167, 172, 187
  moving 86
  offset bolted 68–71, 206
  scarf 65
  screwed 64–6

Kettle 117
Kinematic design 84–6, 95

Lathe 112–13

Leaking and bursting 22, 108–9, 190, 205, 206, 210
Little end 93–4
Load
  alleviation 42
  capacity of bearings 99
  distribution 85–6, 169, 195
  in window stay 199
  sharing in gears 209
Lug 68–71, 155–6

Machining 52, 55, 139, 178
Manufacture 52, 55, 63, 123–7, 211
  design philosophy of 126
  gears 138
  generative 125–7
  hard materials 138
  marginal cost of 125
  'near net form' 126
  of shaft 8
  relation to design 124
  replicative 125–7
Matching 16, 46, 66, 147, 205–6, 214
Materials 123–4
  and manufacture pairs 137
  charts 142–3
  choice of 129–31, 142–3
  cost of 131–2, 139
  density 128–9
  figures of merit for 131–7, 142–3
  in actuator 142, 154
  new 141
  of ship 9
  specific stiffness of 128
  specific strength of 128
  stiffness of 133
Mating surfaces 52–5, 173, 179, 211
  in big end 60–2, 191–2
Maxwell, structural lemma 48–9
Moment, bending 27–31
Moulding 124
  torsion in 141

Nature 139
Nesting and stacking 110–12, 114, 118

O-ring 77–8, 109, 150, 157, 178, 203, 207

Pertinacity 48–51
Plates 31

INDEX 225

Polymer 127, 151
  fibre-reinforced 124, 127, 130, 141
Pressure vessels, thin-walled 33–4, 48–9, 155, 204
Prestressing 101
Princess and the pea 195
Principles 208–19 (*see also* under individual names)
  elastic design 209
  force paths 102–4, 205, 212, 213–14
  functions, separation of, etc. 18–22
  kinematic design 209
  least constraint 83–4, 85, 169, 187, 208, 209
  local closure 74, 205
  nesting order 118, 216–17
  'never make two faces' 192
  uniformity 212
Procrustes 149
Production *see* Manufacture
Pump
  bent-axis 180–5
  Clerk Tri-link 188–90
  gear 107–8
  swash-plate 175–80, 185

Quill-shaft 209

Radius of gyration, k 27–8, 63, 100
Reactor, Advanced Gas-cooled (AGR) 22
Redundancy 80–1
Register 59, 61, 73–4
Relief 54–5, 60–1, 114, 211
Repêchage 14
Repertoire (of means) 24, 25, 86, 187
Restrictor 87–8, 90–1
Reynold's number, Re 90–1
Rolling mills 174
Rotor 71–2, 214–16
Rule, inverse loss 67, 216

Scantling 23–4, 123
Scheme 4
Screw, differential 59
Screwed fastenings 56–9, 62–3
Sealing 29, 70, 77–8, 108–9, 145, 177
  kinds of 77
Section 27–8
  in torsion 32
  modulus of 27–8
  open and closed 32–3
  second moment of 27–8
Self-alignment
  of pinions 168–73
Shaft 81–3
Sharing benefits etc. 42, 216
Shear
  flow 29–30, 33
  in beam 28–30
Ship, materials of 123
Sintering 138
Specification 3, 146
Spigot 73–4
Spring 44–8, 51, 133, 137
  in scales 38–44, 45, 130–1
  in suspension 46–7
  in window stay 203
Stability 25–6, 88, 132–3, 168
Statement of problem 3
Stator blade fixing 62–4
Steam engine 9–10
  Watt and 21–2
Steam turbine 140
Steel 123, 124, 128–9, 130, 199
  austenitic 139
  stainless 134, 145, 150
Steering 111, 115–17
'Sticky drawer' effect 113–14, 152
Stiffness 7, 8, 47, 142–3
  in bolted joints 75–7, 105
  of bearing 91
  of handle 196
  of planet carrier 162
  of structure 51, 113
  specific 141–2
  torsional 32–3
Stool analogy 163, 167–8
Strain gauges 43–4
Strength 24
  specific 128, 141–2
Stress
  allowable 24
  alternating 25
  contact 34–8, 188
  design 24–5, 51, 131
  equivalent 24–5
  hoop 33
  longitudinal 33
  shear 24, 30, 31–2
  torsional 31–3
Structure 23–4, 71
  efficient 51
  of planet carrier 160–7

Strut 25–6, 128, 135
Suitcase handle 195–7
Surface 53–5
Surface table 86
Suspension 46–7, 111, 114–17, 120
  arm 197–8
  BMW front 16–18
  independence of 18–19
  rear 18–21, 137
Suspension bridge 119
Swash-plate pump 175–80, 185
  side thrust in 179–80
Swingletree 169

Table
  kernel 13, 149–50
  morphological 11
  of options 11–13, 149
  surface 86
Tap, domestic 5
Telescope 86, 118
Tent 26
Thermal expansion 63, 139
Tie 25–6, 134
  in torsion 135
Tin-opener 11–13
Torque
  constraint 187
  in actuator 146
  in bolt 57–9
  in carrier 160
  thread-climbing 58
Torroja, E. 100, 169

Torsion 31–3
  moulded forms for 141
Tresca criterion 24, 136
Tri-link pump 188–90
Turbine, wind 16
Typewriter 110–11

Valve
  internal combustion engine 10
  steam engine 10
  variable timing of 217
Valve-plate, hydraulic pump 176
  freedoms of 176–7
Vierendeel truss 119–20
Volvo bent-axis pump 182–5
  ball-retention in 183–4
  casing 184–5
  piston 184

Wasa 192–5
Washer 58
  conical 63–4
Web 28, 166
Werkspoor 191
Window stay 198–203
  ergonomics of 201–2
  secure 202–3
Wind-up in pinion 173
Wood 128

Zinc-based alloy 127
Zwicky, F. 11